Adaptations of Desert Organisms

Edited by J.L. Cloudsley-Thompson

Springer

Berlin
Heidelberg
New York
Barcelona
Budapest
Hong Kong
London
Milan
Paris
Santa Clara
Singapore
Tokyo

Volumes already published

Ecophysiology of the Camelidae and Desert Ruminants
By R. T. Wilson (1989)

Ecophysiology of Desert Arthropods and Reptiles
By J. L. Cloudsley-Thompson (1991)

Plant Nutrients in Desert Environments
By A. Day and K. Ludeke (1993)

Seed Germination in Desert Plants
By Y. Gutterman (1993)

Behavioural Adaptations of Desert Animals
By G. Costa (1995)

Invertebrates in Hot and Cold Arid Environments
By L. Sømme (1995)

Energetics of Desert Invertebrates
By H. Heatwole (1995)

Ecophysiology of Desert Birds
By G. L. Maclean (1996)

Plants of Desert Dunes
By A. Danin (1996)

Avinoam Danin

Plants of Desert Dunes

With 84 Figures

Springer

Professor Dr. AVINOAM DANIN
Department of Evolution, Systematics, and Ecology
The Alexander Silberman Institute of Life Sciences
The Hebrew University of Jerusalem
91904 Jerusalem
Israel

Cover illustration: Photograph by J. L. Cloudsley-Thompson

ISBN-13: 978-3-642-64636-2 e-ISBN-13: 978-3-642-60975-6
DOI: 10.1007/978-3-642-60975-6
Library of Congress Cataloging-in-Publication Data
Danin, Avinoam.
Plants of desert dunes/Avinoam Danin.
p. cm. – (Adaptations of desert organisms)
Includes bibliographical references (p.) and index.
ISBN-13: 978-3-642-64636-2
1. Sand dune plants. 2. Desert plants. 3. Sand dune plants–Adaptation.
4. Desert plants–Adaptation. I. Title. II. Series.
QK938.D9D36 1996
581.5′2652–dc20 95-37101

Softcover reprint of the hardcover 1st edition 1996

Typesetting: Best-set Typesetter Ltd., Hong Kong
SPIN 10124741 31/3137-5 4 3 2 1 0 – Printed on acid-free paper

Acknowledgments

I thank Prof. G. Orshan whose initiation of the "ecomorphological viewpoint" enabled me to acquire an ideological background to my observations, Prof. A. Fahn who helped me in several questions concerning plant anatomy and morphology, Dr. Inka Dor who helped and guided me in the field of cyanobacteria, Dr. Ilana Hernstadt who critically read the chapters concerning mosses, Ms. Na'ava Eisland who assisted the work throughout all the stages, and Mr. M. Dvoracheck for the scanning electron microscopy. Thanks are due to my friend Dr. M.G. Barbour who gave me feedback on the manuscript as it evolved and critically read the entire manuscript; to my friend P. H. Raven who critically read essential parts of the manuscript and encouraged me throughout. I thank my colleagues at the Department of Environmental Horticulture, The University of California, Davis, for their help in various parts of the preparation of the manuscript: Dr. J. Harding, Dr. H. Lieth, Dr. G. Fairfield, Dr. J. H. Richards, Dr. G. Webster; Dr. F. Hrusa for identifying my plant collection from California and for critical reading of the manuscript, Mr. S. P. Rae for identifying moss species from California; Dr. T. Cope for his help in grass nomenclature and discussions at Kew; Dr. Mary Seely, Gobabeb, Namibia for her help in various dune subjects before and during the preparation of the book; Prof. H. Tsoar for his comments on a few physical processes of sand mobility; Prof. A. and Dr. Ann Witztum for their critical reading of several chapters; Mr. P. Grosmann for most drawings, and Dr. Michal Yuval for drawing figure 48. My sincere gratitude to Barak Danin for his help in handling the electronic files of the book and preparing the indices with the computer. I thank the keepers and staff of herbaria of the following institutes listed in an alphabetic order: The University of California at Davis, The Hebrew University of Jerusalem, Jerusalem, Israel, Hamburg, and Kew.

Above all, I wish to thank and dedicate this book to my wife Drora who helped and encouraged me all the years before and during the preparation of the manuscript.

Avinoam Danin

Contents

1 Introduction

Adaptation is defined (Raven et al. 1986) as a peculiarity of structure, physiology, or behavior of an organism that aids in fitting the organism to its particular environment. Darwin (1859) and his followers explained adaptations as the results of natural selection through the processes of evolution. Views on adaptations to the environment in pre-Darwinian times are referred to (Brandon 1990) as the "Creationist account" whereby organisms were designed by God to fit the demands or hazards of their environments. The following may indicate a clear Darwinian view, which had already been conceived many centuries earlier. In the Babylonian Talmud (Shabbath 31, a; written ca. 1500 years ago) there is a story about two men who wanted to make Rabbi Hillel (who lived some 2000 years ago) angry by asking him many difficult questions. One of the questions was "Why are the feet of the Africans wide ?" "My son, you have asked a great question", said he, "Because they live in watery marshes" (translated by I. Epstein 1938).

Many of the morphological and physiological attributes of living organisms appear to be of adaptive value (Fahn and Cutler 1992). Adaptation is considered to be an hereditary adjustment of the organism to the environment (Grant 1963). For general reviews on adaptations the reader is referred to other books (Grant 1963; Conrad 1983; Futuyma 1986; Brandon 1990).

In the present book I do not intend to search for the mechanisms of evolution of plant adaptations to desert dunes or desert sands. I follow the idea that "evolution is almost an inevitable outcome of stress" (Bradshaw and Hardwick 1989). I shall document the ecomorphological features (sensu Orshan 1986b) of plants found in sand deserts as adaptations for life in specific types of habitats. I shall compare my findings with the syndrome of properties found in plants of nondesert sandy habitats, mainly coastal areas. Naturally, many of my examples are from the flora and vegetation of Israel and Sinai, which I know better than those of other areas. Complementary field information was gathered on visits to Namibia, southern Turkey, and California. I also explored the literature as thoroughly as possible.

Plant names are presented as in the most updated floras for the areas where most examples come from. These are Feinbrun-Dothan and Danin (1991) for Israel and Sinai, and Hickman (1993) for California. Many plant names have been changed in the last few decades. In order to ease communication the most commonly used synonyms or names that are not strict taxonomic synonyms but were used frequently in ecologic or phytosociologic literature, are given in parentheses. Terms that may be hard to find in common dictionaries are explained when first mentioned.

2 Sand Deserts of the World

2.1 Distribution of Sand Dunes

Sand transported by wind covers approximately 6% of the continental surface area of the world and 97% of these areas occur in arid zones (Pye and Tsoar 1990). Much of these sandy areas are in "sand seas" or "ergs" – terms used to describe large sand-covered regions. Sandy soils and dunes are very common in many parts of the world's deserts (Fig. 1). Inland active sand dunes of all morphological types indicate aridity (Sarnthein 1978), and one tenth of the continental areas between the latitudes 30°N and 30°S are covered by sand deserts. Sarnthein (1978) concludes that dunes covered almost 50% of the same area ca. 18000 yr ago, supporting the theory which prevails at present that the last ice-age was a dry period in the tropics.

Some 20% of the world's arid zones are covered by eolian (windborne) sand, and they range from 2% in North America to more than 45% in Central Asia (Pye and Tsoar 1990). In the desert areas of Israel (mean annual rainfall <200 mm), sand dunes constitute some 13% of the total area; in the Sinai (mean annual rainfall <100 mm), they cover 21% of the area (Danin 1983); and in the Saharan Desert their cover is 28% (Tsoar 1982).

Mobile sand is an extremely poor habitat for plant growth because it contains very small quantities of fine-grained particles, organic matter, and nutrients, and in areas with low precipitation no plants grow on open sands (Bagnold 1941). In areas with a mean annual rainfall of 30–70 mm, annual plants may grow on the dune but only in relatively wet years. In less extreme environments, with mean annual rainfall exceeding 100 mm, plant cover may be prominent with perennial plants covering the entire area.

2.2 Sand Mobility

2.2.1 The Modes of Sand Movement

Our main interest in this book is plant adaptations to desert dunes. Since dunes accumulate as a result of sand movement, some common terminology concerning sand mobility is necessary. McKee (1982) recognizes the most important modes of sand movement to be saltation or bouncing, creeping, transport in suspension, and avalanching. When a sand grain which is carried

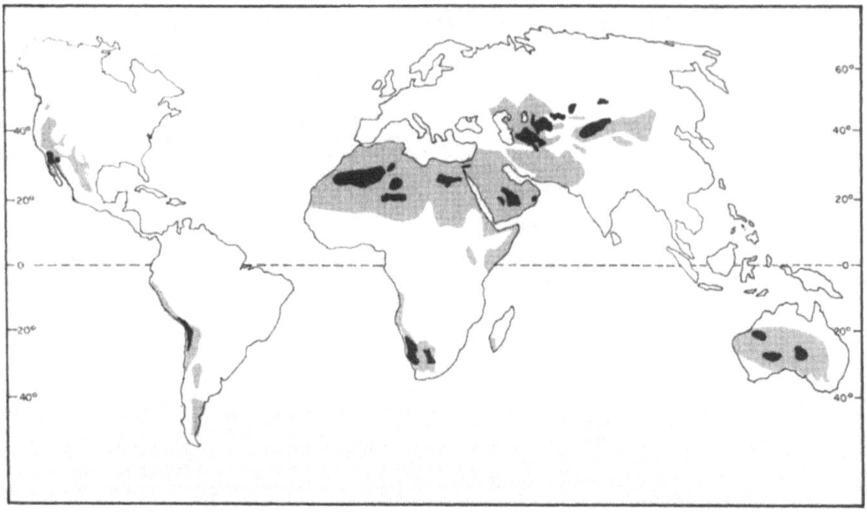

Fig. 1. Distribution of the major desert areas (*grey patches*) and sand deserts (*black patches*) of the world

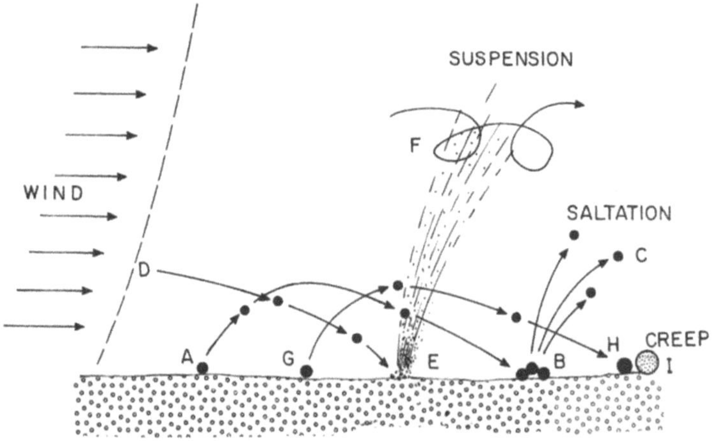

Fig. 2. The ways in which sand grains move. *Letters* represent points for discussion in the text. (After Greely and Arvidson 1990; Tsoar 1982)

by the wind (A in Fig. 2) hits another grain of the same size or a group of grains slightly smaller (B in Fig. 2), they are lifted off the surface to level C, falling and colliding with other grains of the same size and so on; this movement of a few dozen centimeters for each grain is known as saltation or bouncing (McKee 1982; Tsoar 1982; Greeley and Arvidson 1990; Kind 1990). When hitting a grain of larger size (G grain moving to point H and colliding with grain I, in Fig. 2) the latter moves a few millimeters by creeping or by rolling (Greeley and

Arvidson 1990). If a moving sand grain strikes a group of smaller grains (D grain striking E grains causing their lifting off towards F), they go into suspension in the air and may be carried away, even for many kilometers. Some of the components of dust fall recorded on desert margins (Yaalon and Dan 1974) are derived in this way. When wind ceases, the grains in suspension are deposited on the dune surface. If there are plants or microphytes present the dust may be trapped; if the sand is mobile the dust component will continue going into suspension whenever wind is of adequate velocity.

When reaching the lee side or slipface of a dune, the formerly bouncing sand grains accumulate under the influence of two forces. Friction tends to keep the grains still, whereas gravity pulls them down. Consequently the angle of repose at the slipface is about 32°–34° and may be temporarily up to 44° due to the cohesion of wet sand (Fryberger 1991). From time to time, when additional sand or dust grains fall on the sand on the slipface, their weight interrupts the balance, and in consequence slumping, avalanching, or mass movement take place.

There is a minimum wind speed needed to initiate sand movement. The fluid threshold for sand movement is 4.6 m/s and the velocity necessary to produce large-scale sand movement is 10 m/s (Ash and Wasson 1983).

The wind regime in the area of sand deposits influences dune form. Several morphological features of an area covered by dunes at various stages of mobility and stabilization are displayed in Fig. 3. This desert area in N Sinai has several variants of linear dunes. The areas with lightest sand colour are those where the sand moves maximally: area 1 is a patch of four "barchan" dunes; "seif dunes" are marked by 4. Number 3 indicates "braided dunes", as they are named by Tsoar and Møller (1986), and the dark area (area 2 in Fig. 3) among these dunes is covered by vegetation. For more comprehensive accounts of sand mobility and for classification of dune forms, the reader is referred to books or papers dealing specifically with these questions (e.g. Bagnold 1941; McKee 1979, 1982; Ash and Wasson 1983; Greeley and Arvidson 1990; Pye and Tsoar 1990; Kind 1990). A classificatory view of sand movement based on the dynamic reaction of dunes to the wind regime is presented by Thomas (1992). According to his concept, transverse dunes are migratory, linear dunes are forms where mobile sand passes the area, and star dunes are areas of sand accumulation. In each of the three dune types there are sites of sand accretion and sites of sand deflation, and their position and duration have the most important impact on plant life.

2.2.2 Ripples

Ripples are crests formed on the dune surface when sand moves (Bagnold 1941) and therefore they are an indicator of sand mobility. They are formed by accumulation of the coarse grains which move by creeping or rolling. These grains accumulate in somewhat elevated small crests in a position perpendicu-

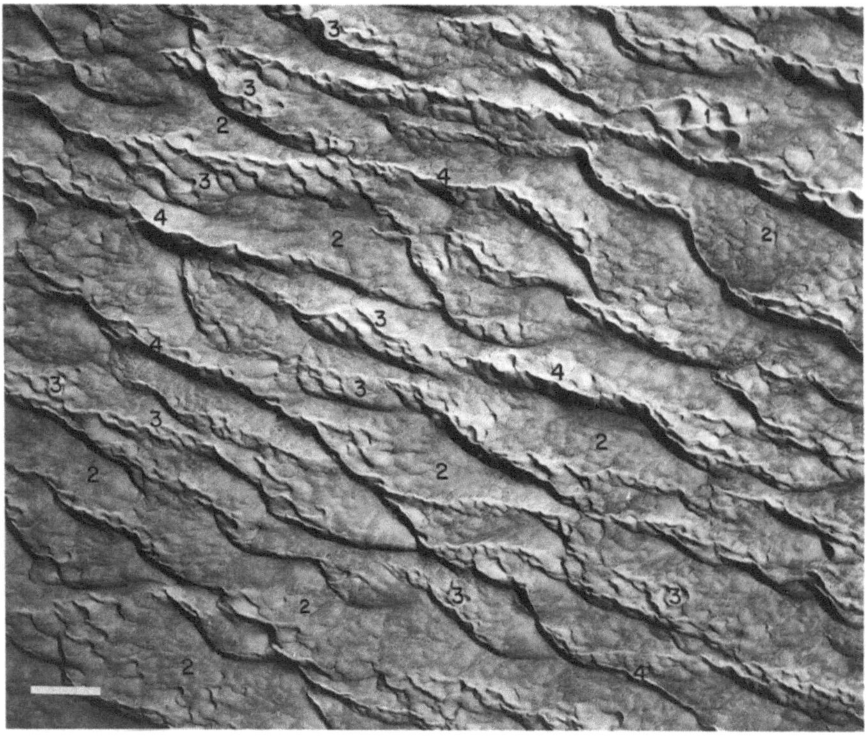

Fig. 3. Aerial photograph of an area with linear dunes in N Sinai. The dark dune slopes and interdune (*2*) areas are vegetated. Barchans (*1*) and dune crests, especially knife-edged seifs (*4*) are devoid of vegetation and are thus white. The linear braided dunes (*3*), typified by small transverse dunelets, represent a complex type of dunes. *Bar* = 200 m

lar to the direction of the last wind event which moved the sand grains. The largest particles, which can be moved only by wind stronger than that which placed them, are at the tops of the crests. This is a state of a microtopographical equilibrium achieved by a certain wind velocity and direction. When these two parameters change, new ripples are formed.

Plants may locally change the wind velocity and direction, and hence the direction of ripples marks in their vicinity. *Elymus farctus* (= *Agropyron junceum*), forming a relatively dense tuft, displays a prominent impact on the ripple direction (Fig. 4). The diversion of wind by *Elymus farctus* causes increased wind velocity in the vicinity of the tuft and hence wind erosion or sand removal are more extensive there. Wind velocity is decreased in the lee of the tuft, and this is where sand accumulation takes place. An area populated by the small coastal grass *Sporobolus pungens* (= *S. arenarius*) (Fig. 5) displays some influence of the plants on the ripples. In addition, each plant has a small shadow dune, the size of which is influenced by the number of stems and the shape of the plant which is the obstacle for air flow. The schematic demonstra-

Fig. 4. Ripples near an *Elymus farctus* tuft on the coastal plain of Israel, after wind has blown from left to right

Fig. 5. Ripples and minute sand mounds in an area populated with *Sporobolus pungens*

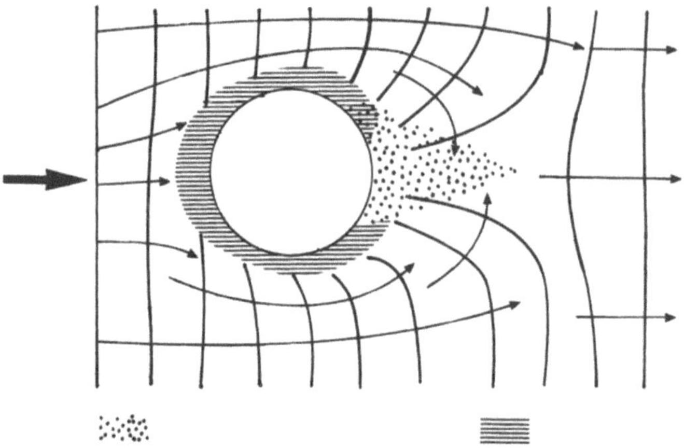

Fig. 6. Ripples and inferred streamlines around a nebka (after Ash and Wasson 1983). *Grains* ⁂ sand accumulation; *lines* ≡ sand removal

tion of plant impact on ripples and the streamlines of wind around its mound in Fig. 6 may help in interpreting the forces which caused the dune morphology of Figs. 4 and 5.

2.2.3 Plants as Modifiers of Local Dune Topography

Plants not only affect ripple direction and size locally, they also cause a substantial topographic change of entire regions. In a schematic representation (Fig. 7), the zones of decreasing wind velocity are shown by areas lighter than the grey background and areas with darker colour indicate zones of increasing wind velocity (Ash and Wasson 1983). The results of prolonged changes in wind velocity in the vicinity of the plant are the formation of shadow dunes (Hesp 1981) and wind erosion near the plant (Figs. 4 and 8). The accumulation of phytogenic hillocks at their leeward side or within the plant canopy was studied by Batanouny and Batanouny (1968, 1969). Such hillocks may vary from a few centimeters to a few meters in diameter and height. Some hillock plants produce adventitious roots and use resources that accumulate in the hillocks, while others remain limited to their original rhizosphere. These plant-caused hillocks are known as "nebkas" (Walter 1973; Walter and Box 1983a; Bendali et al. 1990). Although not included in all dictionaries of biological terms, nebka is a convenient one-word international expression. The same term is used by Cooke and Warren (1973) and by Thomas and Tsoar (1990) who spell it "nebkha". Nebkas have also been called "rebdons" by Cooke and Warren (1973), "shadow dunes" (Hesp 1981), or "coppice dunes" (Thomas and Tsoar 1990).

The influence of plants on wind velocity inside the plant canopy has much to do with the role of the particular species in sand mobility. A spaced or

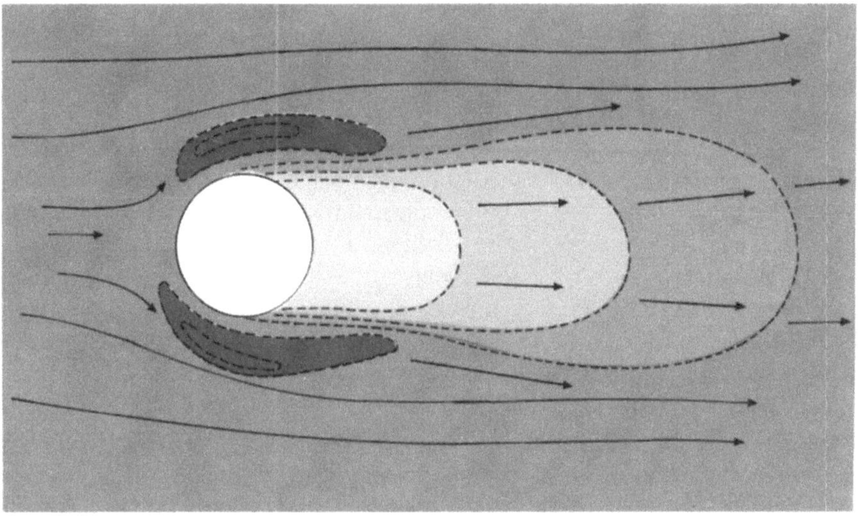

Fig. 7. The influence of plants on wind velocity inside the plant canopy (after Ash and Wasson 1983). The normal wind regime is *grey*, absence of wind inside the plant is indicated by *white*, transition areas are *shades of grey*, and the area with higher wind velocity than normal is *black*

Fig. 8. Dune morphology as influenced by vegetation. *S. scoparia* first built a nebka and then induced erosion near by increasing wind velocity as illustrated in Fig. 7

sufficiently diffuse plant, such as *Stipagrostis scoparia* (= *Aristida scoparia*), decreases wind velocity inside the tuft to a degree that sand accretion becomes possible. It is to be expected that there will be a higher proportion of fine-grained sand inside a nebka than in the sand between nebkas (Danin and Yaalon 1982). The plant continues growing above the sand that accumulates and a nebka is thus formed. As the nebka becomes higher, its impact on wind velocity in its surroundings leads to sedimentation at its leeward side. This impact of a mound on sedimentation is illustrated in Figs. 4 and 6. A growing tuft of a grass may thus ultimately change the entire morphology of the dune (Danin and Nokrian 1991). Other types of plants and their impact are discussed in Section 7.3.1.1.

Nebkas may be formed in the wind shadow of plants which grow in nonsandy terrain but where some sand grains are transported by wind. Plants adapted to the local environment induce the formation of nebkas and, in time, locally increase the percentage of sand in the soil. Plants growing in the gravel plains of Wadi el Ashara west of Fayid, Egypt, trap airborne sand in the wadi (dry water course) channel and enable psammophytes (plants confined to sand) to grow there (Danin 1974; Sect. 5.2.3). In the cold Great Basin deserts of western North America, the halophyte *Sarcobatus vermiculatus* traps sand and builds up nebkas, thus forming local dunes. Such sandy patches may persist for a long time. Whereas the large interdune or internebka areas go through processes of succession and support totally different vegetation, this plant and its locally trapped sand probably survive for millennia. Similar nebkas, formed near halophytes and phreatophytes, are discussed in Sections 7.4.3, 7.4.4, and 7.4.5.

Vegetated linear dunes which become devoid of plants lose their smooth topography giving rise to "braided dunes" and "seif dunes" which can be easily recognized from aerial photographs (Tsoar and Møller 1986). One of the roles of vegetation, according to Thomas and Tsoar (1990), is in determining dune morphology.

2.2.4 Microbiotic Crust as a Sand Fixer

Microbiotic crust develops on the soil surface by the growth of various poikilohydric organisms. When dry, their thalli are dormant but in a latent reversible state. When wetted they rapidly become physiologically active. Cycles of wetting and drying can take place many times without affecting their viability. This crust of a few millimeters at the soil surface has been known by a large number of names. In a recent review of terms by St. Clair and Johansen (1993), the authors imply that "microbiotic" is the most accurate, and their recommendation is followed here. Most of the crust components reproduce by spores or in other ways (but not by seeds). They glue the soil particles together (see Fig. 11; Marthe 1972; Bailey et al. 1973), and thus play an essential role in sand fixation. Crusts include cyanobacteria (blue-green algae), green algae, diatoms (Johansen 1993), fungi, lichens, and mosses. Their general distribution and contribution to sand stability are dealt with here, but the morphology and floristic notes of the various components are discussed in Section 5.11.1.

2.2.4.1 The Succession of Sand Crust Community

Studies of the plant succession on sand dunes near Beer Sheva (Danin 1978b), and the distribution of a cyanobacterial crust in the western Negev (Danin et al. 1989; Danin 1991), reveal the role of the crust in the dune ecosystem. The processes of succession of the crust communities are associated with successional changes in the higher plant communities. These areas have 100–150 mm mean annual rainfall (Rosenan et al. 1985) and approximately 200 nights per year with dew or fog (Danin 1986). Mobile sand, devoid of vegetation, has no crust, and contains up to 1% of fine-grained particles (silt and clay) at the soil surface (Danin 1978b). The first changes in the texture of the dune surface result from a reduction in the local wind intensity caused by the presence of higher plants. By the time that the pioneer community of higher plants has been replaced by the second stage, there is some 2–3% of fine particles at the soil surface, whereas deeper sand layers exhibit the coarse texture of a mobile dune. At this stage occasional filaments of cyanobacteria are found near the soil surface. At later successional stages of both microbiotic and higher plant communities, the percentage of fine particles increases, and the cyanobacteria constitute a prominent crust. Analysis of 20 samples of crust from dunes at various stages of higher plant succession (Danin 1991) revealed a good correlation between the quantity of fine particles and the amount of organic matter in the crust (Fig. 9). The biological meaning of this significant

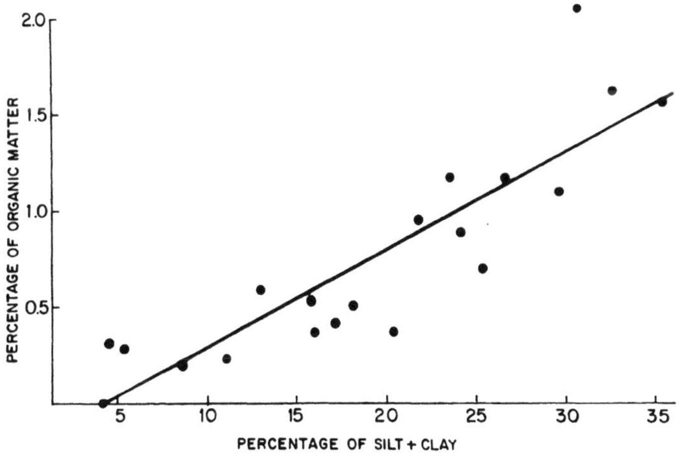

Fig. 9. Scatter diagram of relationships between silt content and organic material in cyanobacterial crusts from the western Negev. Coefficient of correlation r = 0.879; $P < 0.0001$

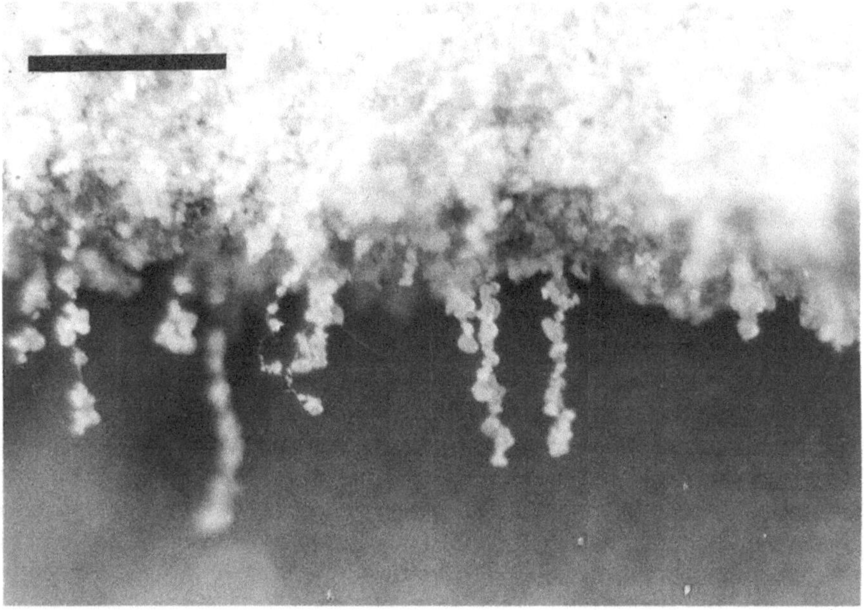

Fig. 10. The lower part of a crust on a stabilizing dune. Sand grains are attached to the filamentous cyanobacteria; *bar* = 1 cm

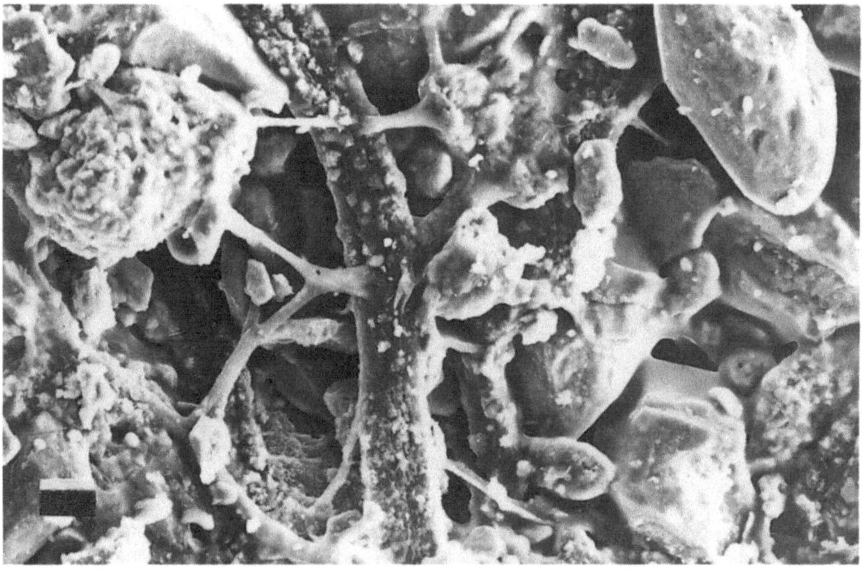

Fig. 11. Soil particles attached to polysaccharid threads from the sheath of filamentous cyanobacteria in a microbiotic crust from sands of the western Negev, north of Nizzana. *Bar* = 10 μm

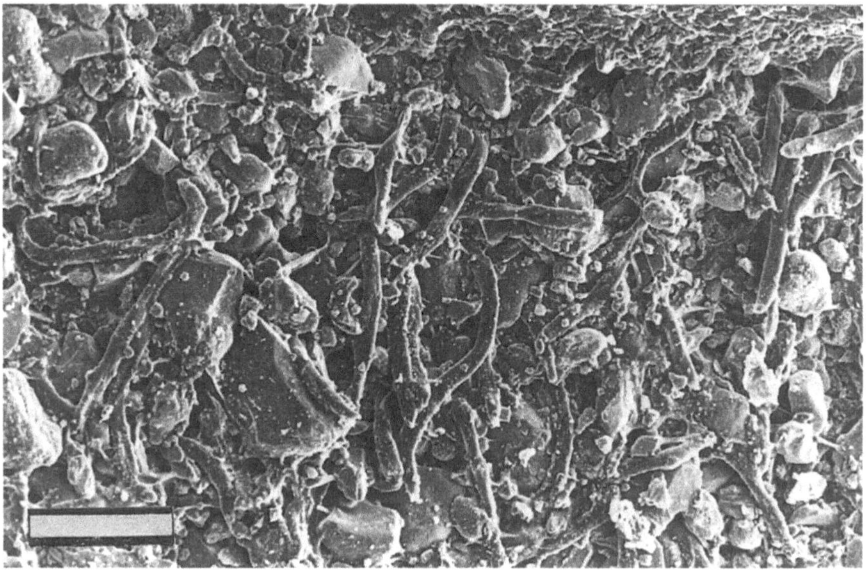

Fig. 12. Microbiotic crust dominated by filamentous cyanobacteria on a sand sheet; *bar* = 100 μm

correlation is that there is a positive feed-back process in the microbiotic crust. After a certain threshold value of fine particles in the soil has been achieved, the cyanobacteria begin to develop. They fix sand and fine particles by gluing their filaments with polysaccharide mucilage to the surrounding grains (Figs. 10, 11, 12). The high moisture-holding capacity of the polysaccharide sheaths of the cyanobacteria and the increasing proportions of fine-grained particles bring about better moisture conditions (Sect. 3.2.2) and enable better growth of microbionts and so on.

Each component of the crust has its characteristic microscopic appearance, enabling the expert to determine each group. Each also has macroscopic characteristics which are easy to recognize.

It is difficult to recognize any change in sand surface structure in the early stages. There comes a stage, however, when filamentous cyanobacteria occur in a sufficient quantity for macroscopic recognition (Fig. 10). The vertical lines of sand grains hanging from the lower side of a cyanobacterial crust are actually cyanobacterial filaments with the sand affixed to them by mucilage (Fig. 11). The filaments shrink when dried and the transparent empty mucilaginous sheaths remain unseen to the naked eye. As a result of the increasing proportions of fine particles, the surface may look different from that of mobile sand. At an advanced stage of cyanobacterial succession, species with thalli which develop with aboveground parts become established. These parts become rich in protective dark coloured pigments (Abeliovich and Shilo 1972) and the formerly white sand becomes grey. When wetted, these aboveground

Fig. 13. LANDSAT image of the border area between SW Israel and NE Sinai (1973). *1* Reworked sand, vegetation and microbiotic crust destroyed; *2* sand, vegetation, and microbiotic crust almost intact; *3* date palm plantations; *4* shifting coastal sands; *5* sandy-loam area covered by orchards; *6* fenced area of Israeli settlements where destruction of vegetation ceased. *Bar* = 20 km. (Courtesy NASA; Danin 1983)

thalli turn black. At this stage the crust surface looks similar to that in Fig. 12 where organisms and inert components of the soil appear as one unit. After cyanobacteria have induced the substantial part of sand fixation, other organisms such as fungi, mosses, and lichens establish themselves and become part of this complicated and poorly investigated component of the desert ecosystem. Much of the reflectance from a sand sheet covered by a well-developed microbiotic crust is in fact from the microorganisms present (Karnieli and Tsoar 1995). The crust is easily recognized in a satellite view (from an elevation of 900 km) of the Earth at the border between Israel and Egypt on the sandy area of the western Negev and NE Sinai (Fig. 13; Danin et al. 1989; Danin 1995a). The sharp line between the areas under different management is further discussed below.

2.2.4.2 Disturbances of the Microbiotic Crust

Due to the role of the microbiotic crust in the sand dune ecosystem, it is important to discuss the processes and agents of disturbance which disrupt this fragile component. There are natural disturbances which are part of the ecosystem and have been functioning for millions of years.

2.2.4.2.1 Natural Disturbances

The most common natural disturbance of the microbiotic crust is being covered by sand. When sand covers the light-demanding organisms they enter their dry latent state. When wetted by rain they receive light signals through the wet sand. Cyanobacteria can perform phototactic movement (Gabai 1985; Dor 1995). How deep the sand can be, yet still allowing light penetration and the upward growth or phototactic movement of cyanobacteria, we have no measure, but Nienow et al.'s (1988) model of light penetration into sandstone rocks in Antarctica shows that wet sandstone has ten times the transmittance of dry rock. Wetting of sand colonized by cyanobacteria and slightly covered by sand leads to positive phototactic movement of the thalli. Leaving their mucilaginous sheaths, they move towards the soil surface. They may then produce new gelatinous sheaths and become situated higher than they were in their pre-wetting position, leading to the stabilization of the loose sand crust. Mosses grow and produce new leaves above the sand and the fine particles covering them (Moore and Scott 1979; Danin and Ganor 1991). This issue is discussed further in Section 5.11.3.

A common local agent of natural destruction of the microbiotic crust, is the activity of burrowing animals. Constant cleaning of burrows and disposal of sand on the crust causes local disturbance. The sand brought up from the deeper layers is of mobile-dune type which is poor in fine-grained particles and devoid of microbiotic components. The growth of the crust community into the disturbed area is gradual throughout the growing season. Growth depends

Fig. 14. A nest of ants in a stable dune 15 km S of Beer Sheva. The ants make small sand balls and dispose of them at the nest entrance

Fig. 15. Burrow of a rodent in an area with well-developed grey coloured microbiotic crust. Note the fresh white sand brought up by the animal

on the size of the disturbed site and the depth of the fresh sand layer, which in turn depends upon the animal's size and the form of the hole that it digs. Examples of animals which dwell in the sand are ants (Fig. 14), lizards, rodents (Fig. 15), and even porcupines (Danin 1978b). Large animals dig to obtain geophyte bulbs, corms, or roots, as well as to excavate burrows. A comprehensive study of porcupines digging bulbs on stony slopes of the Negev Highlands (Guttermann 1987) shows that they make temporary burrows. Such activity in stable dunes causes large disturbances to the microbiotic crust.

Small patches of local crust destruction take place in areas of stable sand where species of *Helianthemum* grow. Many of these have symbiotic relations with fungi, the fruiting bodies of which are tuberous and develop below ground (Sect. 7.3.4.4.3). When reaching their maximal size (up to 10 cm in diameter) they break the crust and may initiate sites of local deterioration and erosion.

2.2.4.2.2 Destruction by Humans and Domestic Animals

Herds of large animals cause local destruction by trampling and crushing the crust. Wind erosion then follows. In many sandy areas, trampling and overgrazing by domestic animals have a profound influence on the plant life (Tsoar and Møller 1986; Danin 1987, 1995a). This is related to the sensitivity of the relatively soft substrate which has a low resistance to penetration by the feet of goats and sheep. If the footprints of domestic animals or man are few and do not touch each other, there is a good chance that the disturbed crust will be restored. When trampling causes fragmentation of the crust which is broken into many small pieces, these can be carried away by winds and thus, be destroyed. As a result, fresh sand is exposed and, being unprotected, reworking may start. This process of fragmentation is not necessarily irreversible. In 1972, a large area of trampled sand in NW Sinai was closed to the Bedouin and their domestic animals. After 2–3 years, the result could be distinguished from satellite images (area 6 in Fig. 13) as a dark-coloured area, much of it due to crust reflectance.

Passing vehicles may cause long-lasting disturbance. The impression left by a single vehicle can remain as a depression in the sand for decades. However, due to regrowth of the crust, the temporary discontinuity becomes unnoticeable within a few years.

2.2.4.3 Distribution of Microbiotic Crust on Sands

The diaspores of organisms inhabiting the microbiotic crust are wind dispersed, and resemble dust. Spores of mosses can survive desiccation from 6 months to 3 years, and some species even up to 10 years; thus, spores of many moss species are capable of long-term dispersal (Zanten 1978, 1984). Spores of many fungi, lichens, and mosses, asexual diaspores of lichens (soredia and isidia) and thallus fragments of cyanobacteria are 1–50 μm in diameter and can be dispersed as aerosols (Akers et al. 1979). These diaspores may be carried for thousands of kilometers. Chatigny et al. (1979) provide data about the dispersal distances of small particles from volcanic eruptions which are capable of changing light infiltration through the atmosphere at distances of thousands of kilometers from the volcano. Particles with a radius of 10 μm or more may be washed out of the air by raindrops or they can fall directly; smaller particles eventually coagulate in clouds and reach the ground in rain drops (Newell 1971). Therefore, every point on earth receives showers of such diaspores. The development of a crust on the soil surface is therefore a function of the fitness of the crust components to the local environment. Microbiotic crust development is a predictable process in dunes where mean annual rainfall exceeds 90 mm, combined with about 200 nights of dewfall or fog (see below, Table 1). In areas where the mean annual rainfall exceeds 200 mm, microbiotic crust develops only if higher plants do not overshadow the surface.The composition of microbiotic crust communities changes according to mean annual rainfall (Sect. 5.11).

Hydration by high air humidity enables some components of the crust, for example *Microcoleus sociatus*, to photosynthesize (Lange et al. 1994a,b). Thus, high air humidity should be regarded as a source of moisture for such poikilohydric organisms living in the microbiotic desert crust. These include lichens (Lange et al. 1994b) and cyanobacteria (Lange et al. 1994a).

Israel is a small country, yet it has a high diversity of climatic and edaphic conditions. It may therefore serve as a model for larger areas. The situation in Israel does not give us complete answers to all questions about the distribution of microbiotic crusts because sandy areas do not occur continuously. However, areas of sand with crust, compared with others where a crust does not occur, may give a clue to understanding large-scale patterns of microbiotic crust distribution. The sands in Israel that experience more than 90 mm mean annual rainfall and a high quantity of dew or fog, develop a microbiotic crust (Table 1) provided that they are protected from trampling and similar disturbance and are not shaded by trees and shrubs.

Where quantities of dew are medium, as in Biqa'at Uvda (indicated by the occurrence of epilithic and endolithic lichens on rocks surrounding the sandy valley), and rain less than 50 mm, no crust develops on the sand. In the Arava

Table 1. Environmental conditions and crust development in parts of the Negev desert

Site	Rain (mm)	Dew and fog	Crust	Restrictions
Coastal Plain	300–500	High	+++	Not shaded
Western Negev	100–200	High	++	
Mishor Yamin	90–100	Medium	+	
Biqaat Uvda	30–50	Medium	–	
Arava Valley	30–50	Low	–	

too, where dew and rainfall are both low, microbiotic crusts do not develop on the sand dunes. Crusts develop, however, in wadis where water accumulates as a result of runoff and on silty ground where the moisture-holding capacity of the soil is higher than in sand. However, the crust is only found in the sections of the wadis where water flows at low energy and does not cause erosion.

3 Environmental Stresses in the Sand Desert

In order to review plant adaptations to sand, one has to focus first on the environmental stresses imposed on plant life. These stresses may be listed under the following headings: (1) substrate stability, (2) moisture regime, (3) effects of nutrients.

3.1 Substrate Stability

Instability or mobility of the ground seems to be the most destructive factor for plant life in sand dunes, and hence the most important environmental factor. The mobility of sand creates a changing level of soil surface in relation to plant parts. High mobility of sand may lead to burial of the entire plant in sites of sand accumulation, or to the exposure of much of the root system in sites of deflation.

3.1.1 Sand Accretion

The sites of sand accretion in dunes of different shape and type are mostly on the lee sides of the dunes. A specific site on a specific dune may change in position when the direction of wind, strong enough to transfer sand grains, changes. Due to a steady wind regime, certain locations within a dune system have a higher chance of becoming accretion sites than others (Tsoar and Møller 1986). For example, seif dunes are formed in desert areas with strong winds essentially from one direction, but with slight changes in direction from winter to summer. In such dunes a certain location will be subjected to accretion for a few months and then to deflation as the seasons change.

Plants on the slipface, where slumping or mass movement take place, are vulnerable to burial by sand. In northern Sinai this is a special habitat (Danin 1983) for a few species which do not grow elsewhere in that area. This habitat is further discussed in Section 6.3.

The interference of plants with sand movement may change the kind or intensity of sand accretion which, in itself, influences plant development. As a result, a sequence of changes occurs in the environment, and hence in the vegetation. The mobility of sand may interfere with germination and seedling establishment. Possessing a mechanism to overcome sand accretion during germination confers a biological advantage on plants growing in such habitats. When plants grow on sand they decrease its mobility, primarily by reducing the wind velocity near the soil surface. Since wind velocity is the main grain-sorter of sand, its decreased intensity increases the proportion of small-sized particles (Danin 1978b; Danin and Yaalon 1982). Such changes lead to changes in the moisture regime and to an increasing potential carrying capacity for vegetation. Plants in nebkas also change the direction and intensity of wind in their wind shadow. Sites behind nebkas may become windless and the margins became sites of deflation (Figs. 7 and 8). In desert plains where some sand is transported by wind, plants growing in a nonsandy substrate may induce sand accretion and the formation of local dunes. Phreatophytes such as *Tamarix aphylla* in the Arava Valley, Israel, and in coastal areas of western Sinai, *Salvadora persica* in SE Sinai, and *Prosopis glandulosa* in southern New Mexico (Campbell 1929) and in Death Valley, California, build up large nebkas. Where such trees grow densely, nebkas meet and form continuous dunes which eventually create a habitat for other plants as well.

3.1.2 Deflation

Sites with poor protection from wind erosion become vulnerable to deflation. Depending on their velocity, winds may remove fine-grained particles as aerosols. As the vegetation cover decreases, the extent of sand removal may reach a level that influences plant life. Exposure of roots has certainly been an important selective factor in the evolution of sand-desert plants. The constant erosion of fine-grained particles creates a coarser texture (Danin and Yaalon 1982; Danin 1983) and causes changes in soil moisture. When dunes become stable and the wind direction changes, areas with constant deflation are formed. Some plants are adapted to these specific environmental conditions (Sect. 5.4.4).

3.1.3 Blowouts

Blowouts are sites of intensive deflation which may be from a few dozens to a few hundreds of meters square. Blowouts are rather common in coastal sandy areas (Watt 1937; Barrere 1992; Jungerius et al. 1992). The direct initiator of the localized wind erosion under certain circumstances is a whirlwind travelling at a relatively high speed and having considerable lifting and carrying powers, similar to a miniature tornado or cyclone (Watt 1937). Early seral communities in coastal dunes of nondesert areas do not have extensive vegetation cover and thus are much more susceptible to blowouts than later stages of succession. Detailed mechanisms of formation and the forms of blowouts in European coastal dunes are discussed by Barrere (1992). In arid lands, the main protection of the ground from blowouts is the microbiotic crust (Danin 1978b; Danin et al. 1989; West 1990). When this crust is affected (Danin 1987; Sect. 2.2.4.2.2), blowouts can easily occur. Most of the area near Nahal Secher, in the N Negev, described by Danin (1978b) was once covered by microbiotic crust. By 1986, as a result of intensive cutting, grazing, and trampling by the Bedouin and their herds, at least 2–3 m of sand had been removed by wind erosion.

3.1.4 Injuries by Airborne Sand

Windborne sand particles, moving by saltation, may injure the aerial parts of plants (Boyce 1954; Pavlik 1985). The abrasive effect of sand has influenced plant evolution in dunes. A protective layer of sand on the plant surface may be of advantage and thus may be considered as an adaptation.

3.2 Moisture Regime

The moisture regime is strongly associated with the texture of sandy soils. Mobile sand is coarse-grained, and composed of fine to coarse sand with a very low proportion of silt and clay (Tsoar 1974; Danin 1978b; Danin and Yaalon 1982; Seely 1991). Such a substrate has a very low water-holding capacity (Kramer 1969). The upper sand layers may become dry soon after rain falls, thus limiting seed germination and seedling establishment. In addition, in dune areas where vegetation cover is minimal, some water is lost to deep drainage (Prill 1968), beyond the reach of plant roots. The driest type of sand, totally devoid of vegetation under extreme desert conditions, because of the nature of its mobility, is the solitary "barchan". These crescent-shaped sand masses may move up to 100 m/year on gravel plains (Walter and Box 1983a). While moving, each sand grain is exposed to the arid atmosphere more than once a year and is dried. The rate of migration of barchans tends to decrease as dune size increases (Norris 1966; Hastenrath 1987). If the rate becomes low enough, one may expect some vegetation on sections of slow moving barchans. Other types of sand dune may also lose water as a result of their mobility.

Stable sand, covered by a microbiotic crust rich in silt and clay (Danin 1978b; Danin et al. 1989), may lose water by runoff (Yair 1990) and direct evaporation from the crust, which has a high water-holding capacity. As a result, the sand beneath the crust becomes deficient in water. The crust may additionally influence the germination of seeds by being a physical obstacle to the penetration of diaspores into the ground, and by interfering with the direct contact of diaspores with the soil particles which transfer the moisture necessary for imbibition and germination.

Little of the water that falls in precipitation is lost. Much is retained in the deeper layers of sand dunes because of rapid penetration, small losses through direct evaporation from the soil surface, and the mulching effect of the dry top sand layer (Chadwick and Dalke 1965; Leistner 1967; Noy-Meir 1973; Zohary 1973; Pavlik 1980; Orshan 1986a; Seely and Louw (1980).

Plants may also obtain water in sandy deserts from moisture condensing near the sand surface. Jackson (1973) found a diurnal variation in the water content of the surface soil under field conditions. Marshall and Holms (1979) further explained these changes as "thermally induced moisture flow; nocturnal cooling by radiation from the ground surface is often observed to cause an early morning accumulation of moisture in a rather thin layer near the surface of the soil".

3.3 Nutrients

The nutrient content of the almost pure quartz deposits in many sandy areas is very low (Ranwell 1972; Buckley et al. 1986). In his study of the nitrogen requirements of dune plants in Australia, Buckley (1983) concluded that the sparse vegetation at the dune crest and the absence of *Triodia basedowii* were linked to the poverty of the soil in nitrogen. Other species of *Triodia* do, however, grow in such nitrogen poor situations (Buckley 1983). The dune flanks in the same study area, which are richer in nitrogen, have a richer vegetation cover. Buckley et al. (1986) later reported that the Tengger dunefield of China contains even less nitrogen than the sand on the crests of central Australian dunes and this is associated with low vegetation cover. The findings of Buckley and his colleagues indicate that even the oligotrophic plants populating dunes differ in their demands for nutrients.

Poverty in available nitrogen is bypassed by perennial grasses which contain N_2 reducing bacteria in their rhizosphere (Wullstein 1991). These grasses have persistent root hairs with sand grains attached by mucilage, the combination of which is known as "rhizosheaths" (Sect. 7.3.4.4.2). The relatively high moisture content of the rhizosheath provides a nanohabitat which fastens the growth of N_2 reducing bacteria (Wullstein 1991). The establishment of pioneer grasses may also lead to a decrease in sand mobility, enabling the establishment of additional nitrogen-fixing organisms. An example of this process is the establishment of cyanobacteria following sand stabilization by *Stipagrostis scoparia* in the western Negev (Danin 1978b; Danin et al. 1989; see also Sects. 2.2.4, 5.11, and 6.1).

4 Habitat Types of Desert Dunes

The wide ecological range of desert dunes should be divided into two climatic types. (1) Dunes of extreme deserts where mean annual precipitation is less than 50 mm. There is insufficient water stored in such dunes to support perennial plants. Perennials growing on these sands (such as stands of *Haloxylon persicum* in the southern Negev or the Arava Valley) use water in the substrates below the sand. Annual plants may briefly cover such dunes in years of sufficient rainfall. Under these conditions, plant impact on dune morphology is mostly negligible. In these extreme desert areas, nonsandy soil types may support vegetation in the contracted pattern (i.e., plants restricted to dry water courses; mode contracté, Monod 1931; Danin 1983). (2) Dunes in deserts with more than 70–90 mm mean annual rainfall do support perennial plants, and these form and modify the dunes. An attempt is made, in Section 8.3, to delimit the boundary between extreme and nonextreme desert areas, using the ecomorphological traits of plants.

4.1 Mobile Dunes

Several main habitats, composing the dune field are recognized (Bagnold 1941; Buckley 1983; Seely 1991; Fig. 12.1): the crest, slipface, plinth or base of the dune, and the interdune area.

4.1.1 Crest

This is the uppermost part of a dune which moves back and forth as a result of daily or seasonal changes in wind direction. Severely desiccated by wind and sometimes moving a distance of 1–2 m per day, this habitat does not support any plants (area 4 in Fig. 3).

4.1.2 Slipface

Sand mobility is not as intense on dune slopes as on the crests. Plants may suffer here from sand moving by saltation, creeping, and rolling, and also from slumping or mass movement typical to the slipface. Naturally, the upper slopes of the dune have higher rates of sand mobility than do the lower parts, as discussed for the Namib grasslands by Yeaton (1988, 1990). Sand mobility in this habitat mainly leads to accretion; however, deflation may occur when wind direction changes.

4.1.3 Dune Base, Plinth

The lower parts of the dune slope have very low rates of sand movement. The slopes in this habitat are too gradual to promote slumping or mass movement, and wind speeds are low. Plant adaptations to intense sand accretion or sand deflation would not confer any biological advantage in this habitat.

4.2 Stable Dunes

Dunes may become stable in desert areas at the conclusion of processes of plant succession. However, human management may influence stability, and the boundary between Israel and Egypt displays stable (area 2 in Fig. 13) versus reworked dunes (area 1 in Fig. 13). Stable dunes recovering from intensive trampling occur in the western Negev (Sect. 6.1). An old and stable dunefield is discussed in Section 6.2. Geomorphological evidence for the past mobility of the dunes are nebkas around some plants, and the coarse-grained texture of the soil profile (excluding the crust which may be rich in fine-grained particles).

In drier areas, stable sand dunes have a layer of coarse sand which protects them from further wind erosion. In northern Sinai, the linear sand dunes are superimposed on an area of older stable dunes which have a coarse sand grit and pebble layer at its surface, and which is populated with plants (area 2 in Fig. 3).

4.3 Sites of Constant Deflation

Large areas in N Sinai which look like stable dunes are actually under a regime of constant deflation, as indicated by the layer of coarse particles on the surface. The roots of most perennial plants are regularly exposed (Danin 1983) and there is a tendency towards increasing proportions of coarse-grained sand or grit and gravel particles at the soil surface, due to their greater resistance to transportation by wind. Old linear dunes, formed on the eastern banks of Nahal Nizzana, northern Negev, some thousands of years ago (Zilberman 1982, 1989), face strong westerly winds and are subject to constant deflation. Such a habitat, with hardly any sedimentation, is distinctive and supports a specific vegetation of plants adapted in various ways to sand deflation.

4.4 Sand Covering Other Substrates

When the sand depth above an already existing soil is less than 5 m, the roots of many species are anchored in the substrate and the sand plays a different role than that described in Section 4.1.

4.4.1 Mobile Sand Covering Sand Sheets

Mobile dunes covering a sand sheet have two principal zones: (1) a barren zone where plants are killed by moving sand; and (2) a vegetated zone at the foot of the dunes where the amount of sand slipping from the dune does not kill plants. A habitat is formed in the ecotone of the dune and sand sheet which supports denser and taller vegetation than is found in the sand sheet or the dune. Plants are constantly covered by sand in this habitat, but they need to have the ability to produce adventitious or shootborne (sensu Barlow 1986) roots. Large individuals, and the high density they achieve, may be interpreted as a consequence of their exploiting larger amounts of water in their rhizosphere than do plants of the sand field. The additional water supply they have is assumed to be from the ground below the dune, charged by percolation as reported by Prill (1968).

4.4.2 Sand Covering Hills or Plains of Hard or Soft Rocks

This is a habitat which supplies larger quantities of water than does pure sand because of the high water-holding capacity of the overlain strata (Danin 1983). Larger quantities of water occur in this habitat than on lithosols because of the high rate of infiltration into the sandy soil and the mulching effect of the sand cover (Danin 1983; Fig. 6). Plants growing here have to survive constant alternating coverage and removal of sand from their aboveground parts.

4.4.3 Sand Covering Salt Marshes

Salt marshes are a common feature of coastal areas and inland depressions in deserts (Danin 1981a). Many hydrohalophytes trap airborne sand grains, moving by saltation, and finer particles moving as aerosols, and thus form nebkas (Fig. 16; Batanouny and Batanouny 1968, 1969; Chapman 1976; Danin 1983; Evenari et al. 1985; Bendali 1987). If sand completely covers the plants, these salt marshes become devoid of any vegetation, as in many sand-covered salt marshes in N Sinai. However, there are places where salt marsh vegetation survives among the dunes whereas the dunes themselves support dune vegetation (Fig. 17).

Fig. 16. Small nebkas at the leeward side of *Halocnemum strobilaceum* plants rooted in a sand-covered salt marsh, N Sinai

Fig. 17. Salt marsh dominated by *Halocnemum strobilaceum* in northern Sinai. The dunes covering its margines are dominated by *Stipagrostis scoparia*

Fig. 18. An immense nebka formed around a grape whose roots are tapping a high water table in northeastern Sinai

4.4.4 Sand Covering Shallow Aquifers of Fresh Water

Sand may move and cover sites with a high water table of fresh water. Such sites are appropriate for the establishment of phreatophytes which may then persist and prevail in the area for a long time. Phreatophytic trees, shrubs, and vines can create nebkas several meters high. Such a situation is illustrated in Fig. 18, where a grape plant (*Vitis vinifera*) at the northeastern coastal area of Sinai has formed an immense nebka.

4.4.5 Sand Covering Dry Water Courses (Wadis, Arroyos)

Plants growing in wadis which do not reach the water table but receive a constant supply of water through deep roots, function as sand traps (see Sect. 2.2.3) in a similar way to the plants mentioned in Sections 4.4.3 and 4.4.4. Depending on the size of the wadi (Danin 1983), small plants or large trees can grow, thus forming nebkas or even local dunes, according to the amount of sand available.

5 Plant Case Histories and Ecomorphological Types

In this chapter, the morphology and autecology of several significant components of the vegetation of sand dunes are discussed. The classification of plants into morphological types is presented according to habitat. In other words, these are neither separate ecological nor morphological types, but ecomorphological types. Whenever necessary and possible, a species has been selected as a typical representative of each ecomorphological type.

The key to symbols used in the diagrams is given in Fig. 19. Not all leaf sheaths and axillary buds are illustrated in the diagrams. Almost all brachyblastic stems (i.e., they have short internodes) have axillary buds, either at the axils of the scales or at the axils of leaf scars.

There is a lack of international agreement on the use of terms for the belowground parts of monocots. The terminology of Du Rietz (1931) is precise but unfortunately not used much in contemporary literature. Du Rietz's terms were recommended by Orshan (1986b) whose system requires precise morphological distinctions. Bell (1991) summed up, in a concise way, many terms regarding the Gramineae, but failed to create some which are necessary for describing all the plants in this book. "Tillers" are lateral branches of the parent shoot, developing from axillary buds (Arber 1934; Bell 1991). When a tiller is trapped between the parent shoot axis and the sheath of its subtending leaf, its growth is termed "intravaginal". The tiller will then grow and terminate distally in a vertical culm and an inflorescence. Such sympodial growth is illustrated in Figs. 20, 30a,b, 31, 35, 36, 38, 53, 54, and 66a–c, and discussed in Sections 5.1.1, 5.2.3, 5.3.1, 5.3.4, 5.6.1, and 5.9.5.

Bell (1991) did not provide a name for the subterranean horizontal stem which develops at the base of sequential tillers, but Du Rietz (1931) applied the term "leafy rhizome", which I too shall use. The terminology used for bamboos (McClure 1966) may be of help. The "leafy rhizome" or "scaly rhizome" of Du Rietz becomes a "pachymorph rhizome" (thick rhizome) for bamboos (McClure 1966; Fig. 2). The narrow base of a pachymorph rhizome is the "neck" of bamboos (Arber 1934; McClure 1966) and is homologous with the section between P and L_4 in Fig. 54a. The section between L_5 and L_{10} in Fig. 54 is homologous with McClure's "rhizome proper". There is a high similarity between the schematic presentation of the entire "pachymorph rhizome system" in McClure's Fig. 2 and that of *Stipagrostis* (= *Aristida*) *ciliata* in Fig. 54. Kukkonen (1994) used the term "stem base" for a part of this leafy rhizome formed by several, mostly short, internodes at the base of each ramet (a unit of vegetatively produced population). The result of compact intravaginal growth,

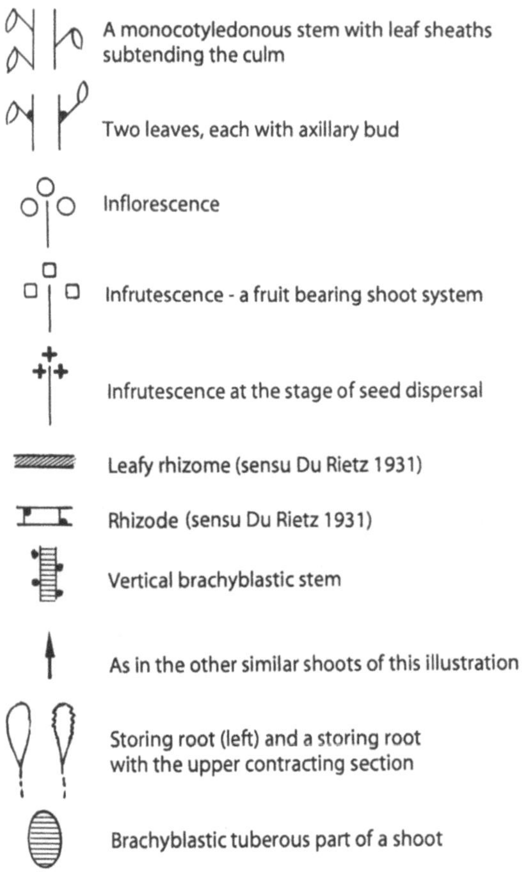

Fig. 19. Key to symbols used in schematic figures

where all the culms come out of a sympodial rhizome, is the 'caespitose' habit. This is known as 'clumping', and is responsible for 'bunch grasses', 'tussock', and 'tufted' grasses (Bell 1991). When a tiller grows horizontally, from the parent shoot, in so doing breaking through the base of its subtending leaf sheath, this is "extravaginal tillering" (Figs. 30c–f and 66d).

When an extravaginal tiller grows belowground, Du Rietz (1931) called it a "rhizode", McClure (1966) a "leptomorph rhizome" (thin rhizome), Bell (1991) a "rhizome", and Kukkonen (1994) a "stolon". Following Du Rietz (1931), "rhizome" is reserved here for "leafy rhizome" or related kinds of subterranean organ. "Rhizode" is used here for a below ground extravaginal tiller. "Stolon" more commonly refers to plagiotropic aboveground root-forming shoots (runners) with elongated internodes. Stolons, in this sense, produce leaves from ortho-tropic aerial shoots. None of the plants discussed in this volume have such stolons and I shall not use the term. The morphology of bamboos cannot be applied to other Gramineae, especially when the differ-

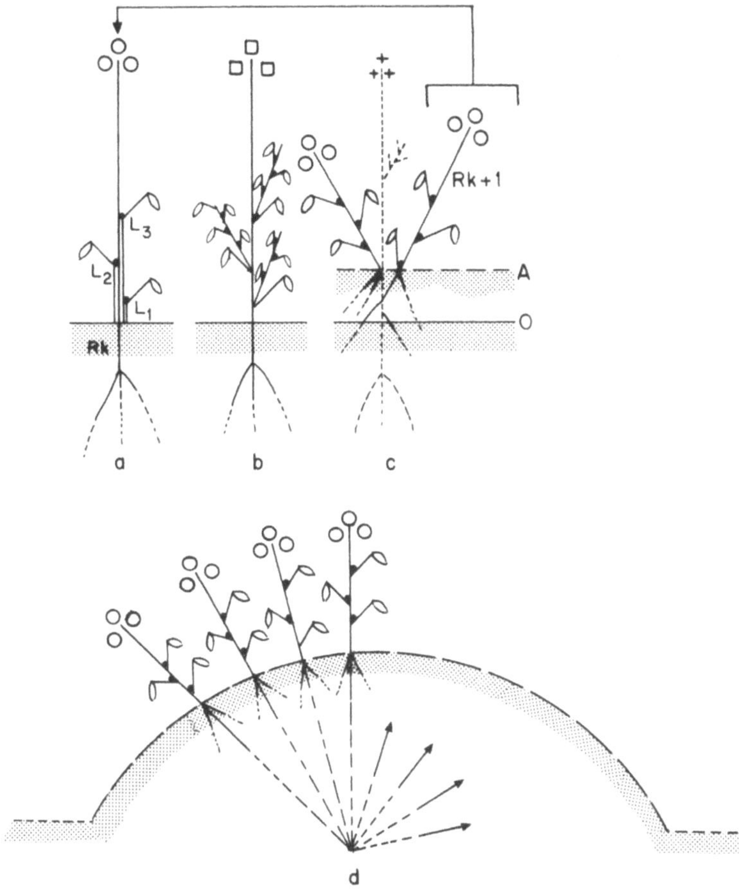

Fig. 20a–d. Schematic presentation of structure and development of *Stipagrostis scoparia*. **a** A single young flowering ramet of the order R_k; **b** the ramet of the order R_k with branches of the order R_{k+1} in axil of aboveground leaves **c** flowering of the R_{k+1} branches and burial of their basal part to level *A* by sand; nodal roots developing from the base of R_{k+1} branch; **d** a nebka formed by ramets (elongating above sand covering them)

ences in rhizome diameter between the thick sympodial "pachymorphic rhizome" and the thin monopodial "leptomorphic rhizome" are not consistent throughout the family. The rhizode of *Panicum turgidum* and *Pennisetum divisum* (Sect. 5.2.3) and of *Lasiurus scindicus* (= *L. hirsutus*) (Sect. 5.2.4) is thicker than the aerial shoots or the sympodial leafy rhizome. "Subolegeocorm" is a term used by Du Rietz (1931) for a rhizode-like subterranean stem which terminates in a stem tuber. It can be applied to species of Cyperaceae. The tuberous part of the cyperaceous rhizome or rhizode carries the leaves on a segment having short internodes. Such an organ is illustrated in Fig. 67d (*Cyperus capitatus*) and discussed for *C. sharonensis* (Sect. 5.9.6).

"Orthogeocorm" is another type of below ground stem which grows vertically and bears roots (Du Rietz 1931). In his definition of this kind of hypogeous stem, Du Rietz comments that these geocorms (below ground stems) are orthotropic or at least not plagiotropic.

"Adventitious root", is a term commonly used in the botanical literature. Various kinds of adventitious roots have been reviewed recently with regard to their origin and terminology (Barlow 1986, 1994; Haissig and Davis 1994). The view of Esau (1977) represents the most common use of the term according to a historical review by Haissig and Davis (1994). Esau (1977) applied the term to roots that arise on the aerial parts of plants, on underground stems, on parts of old root, and in places which have been injured. Fahn (1982) adopted a similar definition. Such a definition is too broad, however, according to Barlow (1986), Groff and Kaplan (1988), Haissig and Davis (1994), and Barlow (1994). Barlow (1986) suggests the term "shootborne root" for roots developing from shoots, "rootborne shoot" to those sprouting from roots, and reserves "adventitious root" for roots regrown in place of lost or destroyed parts or organs (Haissig and Davis 1994). Barlow (1994) developed a system of precise classification for different kinds of adventitious roots, but our main emphasis is on root function and adaptations, so this elaborate nomenclature will not be used in this book. I use the terms "shootborne roots", "rootborne shoots" and "adventitious roots or shoots" synonymously for plant parts that are produced in the context of sand accretion or deflation. The adventitious roots of grasses which develop from nodes are "nodal roots" (Bell 1991). The role of adventitious roots and shoots in the dune ecosystem is very important but poorly understood. In a recent review Haissig and Davis (1994) concluded: "We know almost nothing . . . about the roles of adventitious rooting and roots in the structure and functions of ecosystems of various sorts". Let us define the "reactive growth" of plants, as the kind of growth stimulated by certain events in the habitat. Reactive growth displays, more than many other attributes, the adaptations of a plant to its specific habitat. Convergence among several reactive growth attributes will also be discussed.

5.1 Species Requiring Sand Accumulation – *Stipagrostis scoparia* Type

The members of this group are perennial grasses that are confined to areas of high sand mobility and which require accumulating sand for their survival. They are so well-adapted that their regular life activities are stimulated when they are periodically covered by sand. Their lower stems and nodal roots become senescent before less than 2 years have passed; sand burial stimulates the production of nodal active roots.

5.1.1 *Stipagrostis scoparia*

This species is a perennial chamaephytic grass that is confined to accumulating sand dunes, mostly in desert areas of Asia and Africa. Whereas the base of a shoot is covered with sand it develops nodal roots at the buried nodes (Fig. 20a). Each stem has a short brachyblastic (i.e., with many short internodes) part, ca. 1 cm long, and a much longer dolychoblastic (i.e., with long internodes) part. There is a bud at the axil of each leaf on the dolychoblastic part of the branch. The bud meristems are well-protected by hard scales and by the dry leaf sheath. The bud is situated in a longitudinal depression in the culm, in a position sheltered from desiccating winds. The culm under the leaf sheath is covered by a layer of sticky material. This layer is continuous and stains red with Sudan IV, i.e., it has an oily component. This latter seems to be rather persistent - stems on herbarium specimens collected in the Negev in 1940 are still sticky over 50 years later. I suggest that this material functions to protect the culm and buds from abrasion by sand grains when the leaf sheath has been broken and the culm is exposed. Warming (1909) suggested that a similar mechanism exists for plants growing on sand in Europe.

The basal part of each shoot is brachyblastic, 10–15 mm long and with 4–6 internodes. Each internode in the dolychoblastic upper part of the shoot is 3–4 cm in length. The first leaf in each bud, the prophyllum, is a hyaline scale with scabrous margins. The first regular leaf (L_1 in Fig. 20) has a sheath, a short lamina and a hairy ligule; L_2 looks similar but has a longer lamina; L_3 has a full sized lamina and in most cases has no axillary bud; L_4 and its followers have axillary buds. Shoots usually terminate in an inflorescence if soil moisture conditions are appropriate. After blooming, the axillary buds may sprout and give rise to new tillers (Fig. 20b). When the basal part of such a shoot is covered with sand, 1–4 nodal roots with rhizosheaths develop from the newly buried nodes (Fig. 20c). The roots may be more than 2–3 m long, and many of them run horizontally at a depth of only a few centimeters. The tiller nearest to the new root continues to grow, whereas the old shoot (terminating with an inflorescence) become desiccated and dies (Fig. 20c–d). Each stem may remain alive for 1–1.5 years until it is replaced by new roots and shoots. Green, active, and often blooming specimens of *S. scoparia* may be found in mobile dunes all year round. The ability to be an evergreen fits perfectly its role as a pioneer in sites of sand accretion, because sand movement is unpredictable at any time of the year. In summer 1993, on three locations along the same longitudinal dune in the W Negev, ca. 40 km SW of Beer Sheva, specimens of *S. scoparia* were fully green and blooming where the sand was mobile. In less mobile sand the plants were alive but not blooming and their leaf blades were dry; while a site with stable sand contained dead *S. scoparia* plants.

The plant has no rhizomes or rhizodes. Vertical aerial shoots developed into orthogeocorms or plagiotropic underground organs when covered with sand. Individual tufts of this plant often create nebkas (Figs. 20d and 21). If the

Fig. 21. Nebkas (phytogenic hillocks) around *Stipagrostis scoparia* growing on a linear dune near Har Keren, the Negev

Fig. 22. Nebka created by a tuft of *Stipagrostis scoparia* after many years of sand accretion. When sand mobility ceased, *Artemisia monosperma* became established and *S. scoparia* died. Note the dark cyanobacterial crust. (Danin 1983)

buds on a section of shoots with short internodes sprout, the unearthed stem from which they branch may look like a leafy rhizome. When tillers sprout from a shoot section that has 3–4 cm long internodes, the unearthed underground stem may look like a rhizode. However, since these tillers sprout above ground and only later become covered, they cannot be regarded either as rhizomes or as rhizodes. Vertical stems may be covered when sand accumulates within the center of a loose tuft. Stems at the periphery of the tuft may become prostrate after being covered by sand. In such cases a lateral expansion of the tuft is natural. When not covered by sand, the quantity of green leaves decreases and, when the sand is totally stabilized, the plant dies. I have observed dunes totally covered by dead *S. scoparia* plants in northern Sinai (Danin 1983), and in the northern Negev in a dune stabilized by a plantation of tamarisk trees (Fig. 22; Danin 1978b).

The dispersal unit is a caryopsis, subtended by a glabrous palea and a lemma which terminates in three, nearly equal plumose awns. These are straight when immature, and curve backwards when mature (Fig. 23a). The awn's base is not twisted. The dispersal unit is thus a parachute-like organ and

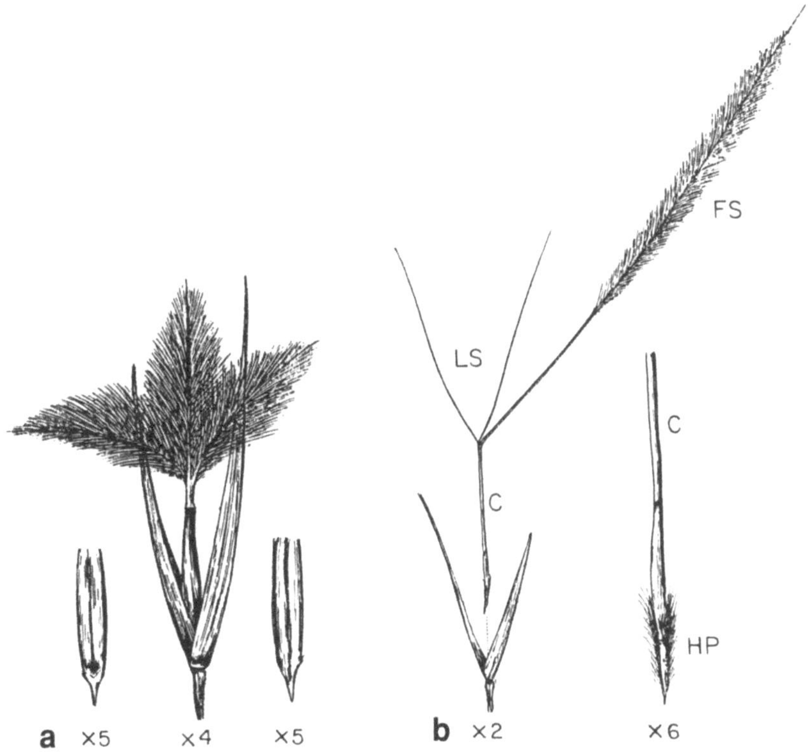

Fig. 23. a A mature diaspore of *Stipagrostis scoparia*: **a** parachute-like apparatus of an awn with three, almost equal, setae. **b** A mature diaspore of *Stipagrostis plumosa* with one setae, being plumose. (Henrard 1929)

looks like a common pappose diaspore of the Asteraceae. During a strong wind in western Sinai in 1975, I saw such diaspores dragged by the wind along the ground surface, rather than carried in the air. They were deposited on the leeward sides of dunes or nebkas where the wind velocity approached zero; diaspores landing there were covered by saltating grains of sand. The ratio of weight to volume and the shape of the diaspores are such that diaspores creep along the surface rather than glide in the air. In this way, the diaspores of *S. scoparia* remain within the system of mobile sand which is the sole habitat of this species. Due to sand mobility, the leeward positions of dunes constantly change, but they are always the most likely sites for the deposition of diaspores of plants dragged by the wind.

5.1.2 *Stipagrostis* (= *Aristida*) *acutiflora*

This species resembles *S. scoparia* in habitat, role in the ecosystem, and architecture. It is a chamaephytic grass which dominates sand dunes in a small area in NW Sinai and in N Africa. The herbarium specimens collected in NW Sinai, 15 km NW from the Mitla Pass in 1975 (HUJ), display the same main features of *S. scoparia* discussed above and illustrated in Fig. 14. *Stipagrostis acutiflora* has leafy stems with hairy culms (*S. scoparia* is glabrous with sticky material on the culm surface) and axillary buds. The adventitious nodal roots are covered with rhizosheaths (see Figs. 34, 39–42; Sect. 7.3.4.4.2). It differs from *S. scoparia* in the structure of its diaspores. The central awn is plumose, straight, and longer than the lateral glabrous and straight awns (Fig. 23b). These diaspores are carried by the wind, and glide high above the ground. The awn's base is slightly twisted, turned 1–2 times, and probably functions as a rudimentary drill. Other species with twisted awns drill the seeds into the ground by the twisting and untwisting of the awn, in accordance with changes of air humidity (Stebbins 1971).

5.1.3 *Stipagrostis* (= *Aristida*) *pungens*

This species is found in southern Tunisia where it occupies mobile sand dunes where accretion takes place (Bendali 1987). *Stipagrostis pungens* was studied by Bendali (1987) and Bendali et al. (1990) who found that it germinates on sand covering stable soil. The seedling decreases wind velocity, induces sand accretion, develops tillers, and forms a tuft. From buried nodes it produces nodal roots which may grow at a rate of 1 cm/day. There is no formation of true rhizomes or rhizodes. Bendali (1987) thinks that the most important adaptations of this plant to sand are its ability to produce adventitious roots and the plasticity of its growth features. Unlike that of *S. scoparia*, the diaspore is armed with a plumose awn and not with a pappose awn.

5.1.4 *Stipagrostis (= Aristida) pennata*

This colonizes the dunes of desert areas of southern Russia, the Caucasus, western Siberia, central Asia, Iran, Turkestan, Pakistan, and Afghanistan (Bor 1970; Leonard 1981). It resembles *S. scoparia* in growth form and dispersal unit. A few subspecific taxa have been recognized. *Stipagrostis* (= *Aristida*) *karelinii*, the dune colonizer and dominant of the Karakum desert (Walter and Box 1983a), is regarded as a variety of *S. pennata* by Henrard (1929).

5.1.5 *Stipagrostis (= Aristida) sabulicola* and Other Southern African Species

Stipagrostis sabulicola is a dune pioneer and mound builder in the Namibian sand sea (Yeaton 1988, 1990). It resembles *S. scoparia* in the location of its renewal buds and the shape of the diaspores (Fig. 23a). It is 1 m high, robust, with thick culms, and stout rhizodes with scale-like leaves (Henrard 1929; Yeaton 1988). The latter properties adapt it to stabilized dunes as well as to mobile ones (see also *Panicum turgidum* and *Pennisetum divisum*; Sect. 5.2.3). Using evidence from mesic dunes of Lake Michigan (Olson 1958a,b), Yeaton (1988) suggests that plants with rhizomatous growth are especially successful at establishing themselves in mobile sands. The fact that many desert dune species of *Stipagrostis* do not have rhizodes shows that rhizodes are not perfect nor are they the only morphologic adaptation in this genus. From Yeaton's (1990) discussion it appears that *S. sabulicola* is better adapted to mobile sand than are *S. lutescens* and *S. seelyae* (= *S. cf namquensis* in Yeaton's papers). In sites where the rate of sand shifting decreases, *S. sabulicola* gives way to *S. lutescens* and *S. seelyae.* In sites where a stable dune slope becomes covered with fresh sand, these latter species are displaced by *S. sabulicola.* They may therefore be included in the types discussed in Section 5.2 or 5.3.

Stipagrostis (= *Aristida*) *amabilis* is confined to mobile dunes of the Kalahari Desert (Leistner 1967). It resembles *S. acutiflora* in its growth form and in the structure of its dispersal unit.

5.1.6 *Swallenia alexandrae*

This monotypic genus is found only in Eureka Valley, Inyo County, California (Pavlik and Barbour 1988). It is a chamaephytic grass which has renewal buds all along its stiff vertical stems. It has no rhizomes and all new shoots develop as intravaginal tillers on older stems when they become covered by sand. The tillers produce nodal roots and become orthogeocorms. Axillary buds are not situated on the node itself, but between nodes. There are no obvious buds in each axil; long stems exposed by deflation show a sequence of 2–5 internodes with bud scars and 2–8 internodes with no sign of buds. The outer scale of the bud is covered with long hairs, as illustrated for *Stipagrostis ciliata* (Fig. 55 Pa and Pb). There is a ring of long hairs covering each node at the outer face, below the leaf base. In this way the bud is entirely covered by hairs. Although the leaf sheaths cover the hairs the latter may function in protection when the leaves senesce and fall.

Under a local regime of sand accretion, individual plants show higher vitality than do plants where sand deflation is taking place. The leaves are stiff and have a thick epidermis. They are covered with short appressed (i.e., pressed flat against another organ) hairs, are rich in sclerenchyma, and have spinescent tips. Even when the whole leaf is dry, the culm remains green and lives for some time. The culm, totally covered by leaf sheaths, may later become exposed as a result of sheath weathering or disintegration resulting from sand abrasion.

According to Hickman (1993), the caryopsis is dispersed by falling from the floret. I have observed that the caryopsis can remain within the hairy palea and lemma after the floret has separated from the mother plant. Either in this configuration or as a naked caryopsis, the diaspore is heavy. It is dispersed by rolling along the surface. In this way, it remains in the dune habitat. When covered by wet sand, the plant produces nodal roots which leave typical circular scars on the dead stem. After a plant has been totally exposed, the long oval bud scars in the center of the internodes and the small round scars of nodal roots can help in reconstructing its history.

5.1.7 *Panicum urvilleanum*

This a robust stiff perennial grass. It is the dominant colonizer of mobile dunes in the Mojave Desert of southern California, including the Kelso Dunes and patches of sand accretion along the Mojave River. It has long rhizodes and short vertical culms which produce nodal roots and thus fit Du Rietz's (1931) definition of "orthogeocorms". It produces monopodial horizontal rhizodes at varying depths (depending on local site conditions), along which every axillary bud is capable of developing, by extravaginal growth, into a vertical tiller. The basal parts of the tillers, close to the rhizode, have 1.5–8 cm long internodes, scale-covered buds, and occasional nodal roots. This basal region is the dolychoblastic part of the tiller. A ring of hairs occurs at each node of tillers and rhizodes. The only function I can see to the hairiness of below ground parts is in case of deflation and exposure. Nodal roots with well-developed rhizosheaths develop from tillers and rhizodes. The leafy part of the tiller is brachyblastic. It starts close to the present soil surface, has 0.1–3 cm long internodes, and is either depauperate or has no obvious buds. Old, long orthogeocorms display alternate dolycoblastic and brachyblastic sections. Each dolycoblastic section above a brachyblastic section implies reactive growth after sand accretion, thus bringing the top of the orthogeocorm close to the new sand surface. Internodes of leaf-carrying sections are 0.2–5 cm long in sites with sand accretion, and 0.1–0.2 cm long in sites of deflation or sand stability. Leaf sheaths are hairy and may vary considerably in length. Leaf sheaths of a plant in an area with modest sand accretion were 5–15 cm long whereas sheaths of plants from leeward slopes with sand accretion were 10–40 cm long. The tiller terminates in an inflorescence. One of the buds on the tiller, on the continuation of the rhizode, or on new rhizodes, will give rise to new tillers. Sometimes the plant looks like a loose tuft and, in this form, it does not form a nebka. At others it forms dense patches which resemble a lawn, a few centimeters above the sand surface.

Panicum urvilleanum has two types of rhizodes. The more common is circular in cross-section, but there is a thicker, flatter type (as if the rhizodes had been subjected to vertical pressure). I could not find any reason for the morphological difference between the two kinds. Exposed rhizodes 2–5 m long were observed in Kelso Dunes. Lateral spreading seems to be dependent on the growth of rhizodes. The structure of the tuft or patch is much influenced by the direction of sand movement. The extent of vertical growth of dolychoblastic tillers seems to be the main mode of growth in reaction to sand accretion. Leaf-sheath elongation is an additional important mode of reactive growth. Exposure of ramets leads to their desiccation and death. The diaspore is a pair of spikelets. Relatively heavy, it is totally covered with long hairs which inhibit gliding and, instead, ensure surface rolling dispersal (Sect. 7.3.5.2).

5.1.8 Summary

The seven representatives of this ecomorphological type withstand sand accretion and tend to be replaced by other plants as the sand becomes stable. These species are components of different genera and floras, and the attributes they share are convergent adaptations. These common attributes are:

1. Taxa in the Gramineae.
2. Stems have buds at the axil of every leaf. Intravaginal tillering takes place above the ground, making them chamaephytes (sand accretion may cause confusion as to where the tillers develop). *Panicum urvilleanum* differs by having rhizodes with the extravaginal tillering of a geophyte.
3. All the plants in this group have orthogeocorms. Their vertical aerial stems may become covered with sand, produce nodal roots, and then emerge again.
4. Nodal roots develop shortly after the node has been covered by sand. The roots produce rhizosheaths. When much of the rhizosphere is exposed for a few days or weeks, that part of the plant dies.
5. They are active all year round and are thus capable of fast growth in reaction to unpredictable sand movement.
6. All the species have buds protected by hairy, rigid scales. Such cover may function as a protection from desiccation, from solar radiation, and from

abrasion by airborne sand. *Panicum urvilleanum* even has hairs on it parts that lie beneath the surface.

7. The stems and leaves have a thick epidermis and are rich in sclerenchyma. These seem to be adaptations that withstand abrasion by airborne sand. A rare adaptation to this component of the habitat is the presence of mucilaginous material on the stems of *Stipagrostis scoparia.* The hairs on the stems or leaf sheaths of *Stipagrostis acutiflora, Panicum urvilleanum,* and other species may be regarded as adaptations to the same factor.
8. Seed dispersal of most species in this group can be regarded as having "directed wind dispersal". All species of *Stipagrostis* with three equal hairy awns, which form a parachute-like apparatus, are dragged along the ground and are confined to mobile desert dunes. American members of this group have rolling diaspores. Both modes of dispersal enable most diaspores to remain in the sand dunes, the habitat to which these species are best adapted. *Swallenia alexandrae* which is endemic to a single dune, has lost nearly all dispersability.

5.2 Species Resistant to Deep Sand Cover or Removal – *Calligonum comosum* Type

Plants of this group are mostly shrubs or tall grasses able to branch from aboveground buds and produce adventitious roots and shoots from buried stems. They can also withstand deflating and stable habitats.

5.2.1 *Calligonum comosum*

This species is a Middle Eastern shrub up to 2 m tall which is confined to sandy areas where the mean annual rainfall is less than 100 mm. It may grow on mobile sand dunes, stabilized dunes, sand sheets, and in sandy wadis. The seedling has small leaves and green stems (Fig. 24a). The adult plant has lignified stems with two kinds of branches – "vegetative" and "generative" (Danin 1964). The vegetative branches are zig-zag shaped (v in Fig. 24 and Fig. 25), green when young, and have thick or swollen nodes. They may grow more than 50 cm in length in one season. Under the bark, at each node, are at least five buds arranged in two groups (Fig. 26). These are accessory buds situated at the axil of the remnant of the leaf of each node (Danin 1964). Similar buds in *Calligonum rubens* are illustrated in Walter and Box (1983a; Fig. 6.12). In *C. comosum,* these vegetative branches later become lignified, brown, and shiny, and constitute the skeleton of the shrub. The lignified body also functions as a large reservoir of water and storage materials by having living fibers (Fahn and Sarnat 1963).

"Generative" branches develop from the accessory buds on the vegetative ones after they have been dormant one year or more. These branches appear to

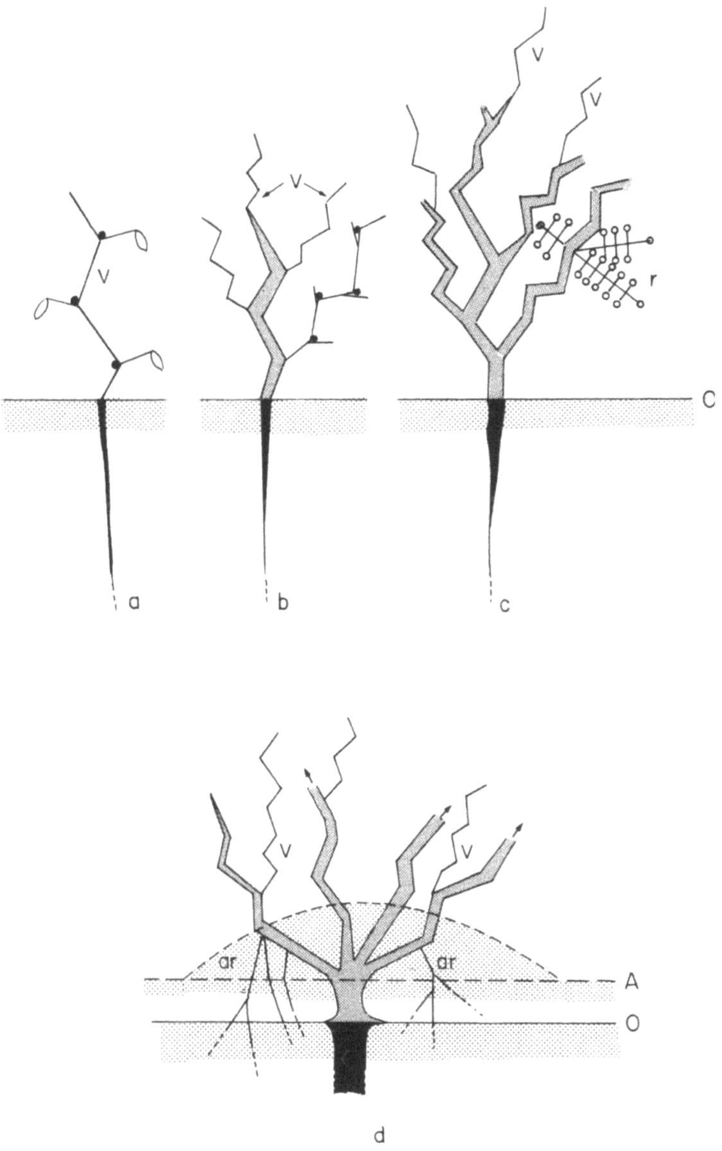

Fig. 24a–d. Schematic presentation of structure and development of *Calligonum comosum*. **a** A seedling with leaves; **b** a young plant with a lignified stem and four zigzag-shaped vegetative branches (*v*); **c** vegetative and reproductive branches (*r*); **d** a nebka where adventitious roots (*ar*) have developed from the plagiotropic part of old lignified branches; new vegetative branches sprout above the ground close to *ar*, *A*-level marks the new level after sand accretion took place covering the *0* level

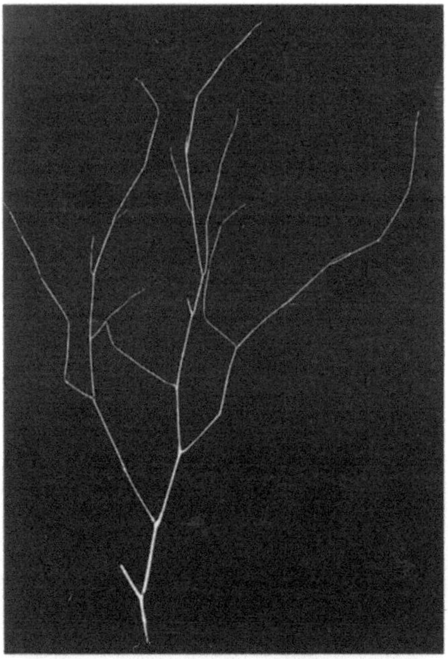

Fig. 25. A "zigzag"-shaped vegetative stem of *Calligonum comosum*

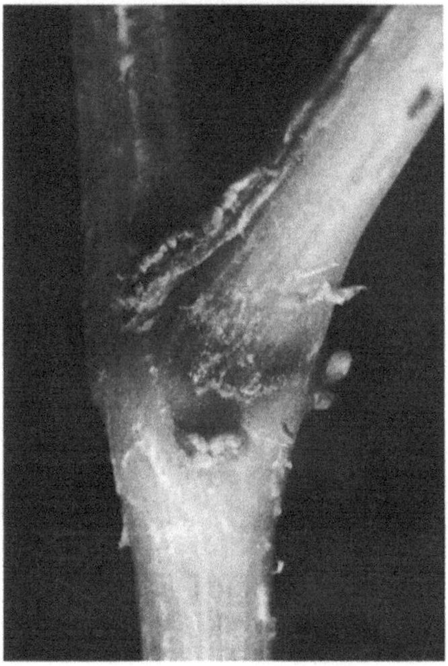

Fig. 26. Accessory buds at the nodal area of a vegetative stem of *C. comosum*

Fig. 27. Generative branches of *C. comosum* arranged in two groups at each node as a result of the sprouting of accessory buds

be whorled due to the proximity of buds in the nodal area (r in Figs. 24 and 27). The generative branches are straight, never become lignified, grow only 30 cm long, and have buds that give rise to branches at their base on the vegetative branches but not in their nodes. However, they bear flowering buds which later become flowers and fruits. At the end of the growing season, towards the beginning of summer, the generative stems are shed. The plant passes the dry season with only lignified stems and is considered to be summer-deciduous. Shrubs planted in sand along the Mediterranean coast near Caesarea, Israel, shed their green temporary stems much later than those in the desert. About a week after the first effective rains generative stems begin to sprout, and a month later they bloom. A microscopic cross section of a green stem is illustrated in Evenari et al. (1982).

The root system of two shrubs was studied 10 km SE of Dimona (Danin 1964). It consisted of a tap root, which penetrated to a depth of more than 1.5 m, and superficial lateral roots which ran horizontally 10–20 m from the tap root. The diameters of vessels in the xylem of the horizontal roots was larger than that of the tap root - Zohary and Fahn (1952) reported similar observations for *Retama raetam* from the Mediterranean coast. When exposed by deflation (Fig. 28), the thick roots withstand exposure for many months without declining.

Fig. 28. A shrub of *Calligonum comosum* after being subjected to sand deflation for a long time

In addition to the main root system, the plant is capable of producing shootborne roots from stems that are covered with sand, even if they are thick and old (ar in Fig. 24d). Roots are produced adventitiously on sand-covered stems and so are young branches. These are produced just above the new roots. In this way, an old plant moves itself by means of younger axes to the periphery of the old plant.

Calligonum comosum builds up nebkas and gradually displaces itself, following the rising sand surface. Having a relatively open crown, it is able to bring its canopy above the developing nebka. It does not need a permanent supply of sand for its survival. When the mother plant becomes old and dies, those peripheral branches that have well developed shootborne roots keep growing and function as independent plants. As a result, clonal circles 5 m in diameter or ellipses 11 × 5 m develop as on the Yamin–Rotem Plain, 10 km south of Dimona, Israel.

The fruit of *C. comosum* is a woody achene with 10–16 rows of bristles (Fig. 84a). According to Koller (1956), germination is inhibited by high humidity near the diaspore. The bristles create an "aeration channel" in the sandy soil around the diaspore. They may also play a role in dispersal. The embryonic root is light sensitive and thus only diaspores covered with sand germinate. Temperatures must be between 15 and 30°C (optimal germination 20–26°C). Germination occurs only in relatively wet years and establishment follows only on sites with low competition.

Ephedra aphylla (= *E. alte*) is a plant which closely resembles *C. comosum*, and grows with it near Dimona. It sometimes grows on a *Calligonum* nebka. It is also a stem-assimilant (having green stems and either diminished or no leaf area; Lyshede 1977, 1979), produces shootborne roots, and can withstand long periods of root exposure. A few *Ephedra aphylla* specimens studied in the coastal area of Israel near Caesarea had rhizode-like organs. Unlike *Calligonum comosum*, *E. aphylla* has diaspores dispersed by birds; the fleshy cones are eaten and the hard-coated seeds excreted later. *E. alata* is another species of *Ephedra* which is confined to desert sands and dominates sandy areas in Sinai (Danin 1983). It has heavy winged diaspores which may be dispersed by rolling (as are winged species of *Calligonum*, Sect. 7.3.5.2). In the Karakum desert, *Ephedra strobilacea* (Walter and Box 1983a) grows in sand dunes, produces nebkas, reproduces vegetatively by rhizodes or stolons (or both - the literature is not clear) and may constitute large patches.

5.2.2 *Artemisia monosperma*

This is one of the most important shrubs of semi-stable and stable sands in desert and nondesert areas of the Middle East. In Israel and northeastern Sinai it is dominant in the mid-seral stages of plant succession, a process by which sand is stabilized (Danin 1978b; Danin and Yaalon 1982; Sect. 6.1). It needs light for germination (Koller et al. 1964) and germinates only in sites where there is a thin cover of sand. These conditions are found in the northern Negev and northern Sinai, between nebkas of *S. scoparia* where wind velocity is decreased. In drier areas, such as the southern Negev, there are other microhabitats fitting its requirements for germination. *A. monosperma* has a denser crown and decreases wind velocity more efficiently than does *S. scoparia*, creating an accumulation of fine sand around itself (Danin and Yaalon 1982). It has the ability to develop adventitious roots from sand-covered stems (Fig. 29). In winter, when the sand covering the stem is wet, shootborne roots may develop within a few weeks of being covered. Stems are able to grow above the accumulating sand, leading to the formation of nebkas 1–10 m long and a few meters wide. In many sites, all the nebkas are derived from one genet (a population unit arising from a single seed) which produces several independent ramets, each with its own root system. In time, the establishment of new *A. monosperma* plants among already existing phytogenic mounds of the same plant leads to the formation of stable sand sheets where sand mobility is minimal (Danin 1978b; Danin and Yaalon 1982; Danin 1995a). Eolian silt and clay are trapped in these sites, improving both the water regime and soil stability.

Artemisia monosperma tolerates sand deflation because it has lignified roots with a corky bark (Fahn and Cutler 1992), thus enabling transport activity to take place in the root, even when exposed. I have seen sites where the exposed root of the shrub is 1–2 m long. An additional morphological adapta-

Fig. 29. Adventitious roots on stems of *Artemisia monosperma* after being buried for a few weeks

tion, probably to sand mobility, is the presence of hairs covering the seedlings of this plant. In the northeastern part of Biqa'at Uvda, Israel, massive winter germination of *A. monosperma* allowed me to observe seedlings in March 1994. Cotyledons, stem, and the first 10 leaves were densely covered by appressed hairs. Leaves developing later from buds on lignified stems are much less hairy or completely glabrous. They are sticky at their bases. The leaf axils of the seedlings are not sticky at all. I assume that the function of hairs in the seedling stage is to provide protection from sand abrasion. When the plant grows older, the function of protecting the sensitive parts from sand abrasion is achieved by the sticky material in the leaf axil. This plant has the ability to produce minute simple leaves, 1–2 mm long, in summer or under drought conditions, and much longer and wider leaves, to 30 mm long under moister conditions. Such changes in leaf size are a common adaptation to drought and are also found in many Mediterranean shrubs and semishrubs (Keshet et al. 1990).

5.2.3 *Pennisetum divisum* (= *P. dichotomum*) and *Panicum turgidum*

In general morphology the two species resemble each other. Both are perennial grasses, 40–150 cm tall, and have glabrous, indurated stems. For much of the year they are leafless and function as stem-assimilants. Both are tufted bunch grasses and have axillary buds throughout the shoots (Fig. 30a). Both may give rise to aerial shoots from stout underground rhizodes and thus fall into Orshan's (1986b) definition of amphiphytes.

Both species have three alternative ways of sprouting and expanding their tufts which may reach 1–3 m in diameter. Most spreading is via scaly stout rhizodes; these extravaginal tillers sprout horizontally and have sharp ends (rh in Fig. 30c). Rhizode diameter is 8–10 mm, and there is a bud at the axil of each scale. These buds may give rise to vertical aerial shoots (R_{k+2} in Fig. 30d). The diameter of these aerial shoots is 1.5–2.5 mm. Expansion can also occur at the base of tillers by intravaginal sprouting of tillers which have become covered with sand (Fig. 30b). Since the aerial shoots have buds in the axils of each leaf, when an aerial shoot is partially covered (and reaches level A in Fig. 30e) it may give rise to new tillers from the buried buds, and roots from the buried nodes. In the course of time, fresh rhizodes develop at the appropriate depth below the new soil surface (Fig. 30f). Although well-adapted to sand mobility, these are not generally plants of mobile sand. In a study site of dune succession in the northern Negev (Danin 1978b), *P. divisum* is confined to a dune which long ago ceased moving.

The roots of both species are covered with rhizosheaths and are sometimes exposed as a result of deflation. In the Sudan, Dr. G. Wickens (pers. comm.) saw an area dominated by *P. turgidum* where much of the space between the tufts was eroded and the living plants were atop columnar structures of sandy soil. On the other hand, in the gravel plains west of Bitter Lake in Egypt, *P. turgidum* participates in a positive feed-back process of building up sand vegetation (Danin 1974). The somewhat sandy ground of the small wadis supports *P. turgidum* which traps sand and builds up nebkas, thus enhancing further expansion of the tufts and the establishment of new plants with their nebkas. As a result, small wadis in the catchment area of Wadi el Ashara dominated by *P. turgidum* are much wider than similar wadis colonized by other plants. Furthermore, the ground of the wadis covered by *P. turgidum* is much more sandy. *P. turgidum* is dominant in vast areas of sand sheets in northern Sinai (Danin 1983).

The diaspore of *P. divisum* is a single-seeded spikelet subtended by setae which assist in wind dispersal. The diaspore of *Panicum turgidum* is a simple caryopsis with subtending scaly glumes. Diaspores of both species are probably dispersed by rolling on the ground rather than by gliding in the air.

5.2.4 *Lasiurus scindicus* (= *L. hirsutus*)

This is an amphiphytic perennial plant (as discussed in Sect. 5.2.3), 40–100 cm tall, with hairy, indurated stems. It may remain leafless for a large part of the

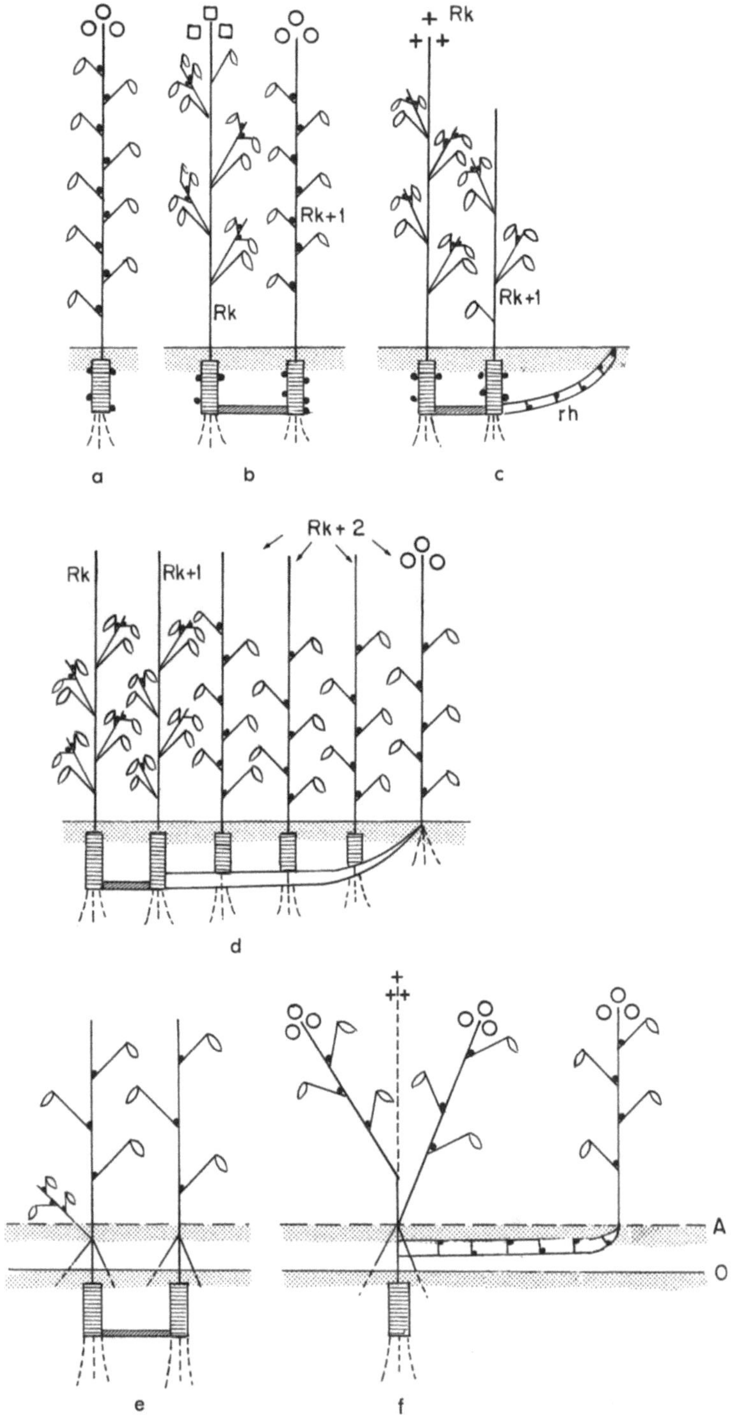

Fig. 30a–f. Schematic presentation of structure and development of *Pennisetum divisum*. **a** A flowering ramet of the R_k order with a bud at the axil of all leaves; **b** R_k ramet with a few branches and a new ramet (R_{k+1}) developed from a bud on the subterranean brachyblastic part of R_k; **c** as b with a rhizode (*rh*); **d** branches of the order R_{k+2} sprouting from the rhizode; **e** and **f** sand cover (to level *A*) and plant response

year and function as a stem-assimilant. It is tufted and has axillary buds throughout its shoots, from the axil of the third scaly leaf on the brachyblastic part of the stem up to the axil of the leaf just below the inflorescence (as in Fig. 30a). It may also give rise to aerial shoots from stout rhizodes. Tillers sprouting from the buds close to the ground tend to have a swollen tuberose base which is rich in parenchyma, bundles of vessels, and fibers. It may function as a storage organ. The bases of shoots developing from aboveground buds are not swollen. The basal brachyblastic part of the tiller is only 10 mm long but it has 4–5 nodes. In contrast, internodes of the dolychoblastic section are 8–50 mm long. The first three scales of each bud are hairy, matching the hairy culm. Lower buds are short and wide, whereas the upper ones are long and narrow. The roots are typically covered with rhizosheaths. The diaspore is a hairy, barrel-shaped segment of the cylindrical inflorescence. It is probably dispersed by rolling on the ground.

5.2.5 *Hilaria rigida*

This is a rigid, hairy perennial grass confined to sand dunes in southwestern USA (Munz and Keck 1963). In the Kelso Dunes, where I studied it, this grass dominates plinth vegetation which experiences the least sand movement. The below ground system of the plant is composed of short sympodial rhizodes, 10–20 cm long and with 2–5 cm long internodes; occasionally there are sections of 2–4 short internodes with axillary buds and well-developed nodal roots. The roots are covered with rhizosheaths. The underground rhizodes are not pubescent, but the outer scales of their axillary buds are. Each rhizode terminates in an aboveground vertical stem which carries an inflorescence. The culm is densely covered with crisp hairs which turn it white. When the leaves dry out, the culms remain green, and the plants function as stem-assimilants. The hairs decrease the impact of solar radiation by reflectance, and may also protect the culm's epidermis from abrasion by airborne sand grains. The culm carries buds in the axils of its leaves. The basal buds are larger than the upper ones. The outer scale of the bud, the prophyllum, is rigid and covered by dense hairs. Buds on the subrterranean part of the stem sprout and give rise to a group of a few vertical hairy tillers, thus forming a loose tuft.

The diaspore is a rounded group of three spikelets with only one fertile caryopsis subtended by many rigid and hairy components (paleas, lemmas, and glumes). The diaspore appears to be dispersed by rolling on the dune surface.

5.2.6 *Tetradymia tetrameres*

This is a rare shrub of sands and sand dunes, mainly in the Great Basin, California (Pavlik 1985). I studied it in September 1994 at old northwestern lake terraces near Mono Lake, California. It is a stem-assimilant shrub, densely covered with white crisp hairs. I found it growing in the least stable parts of a stable dune. When covered by sand, its lignified stems produce adventitious roots, and it forms large nebkas. New branches develop just above the new shootborne roots. In the vicinity of *Tetradymia* were shrubs of *Artemisia tridentata* and *Chrysothamnus nauseosus*. The two latter also form nebkas, and have shootborne roots. They share with *Tetradymia* many adaptations to sand.

5.2.7 *Psorothamnus polyadenius* (= *Dalea polyadenia*)

This is a rare stem-assimilant shrub of desert dunes in southwestern USA. I observed this species in the Eureka Dunes in September 1994. By that time of year there were hardly any leaves on the stems but there were a few flowering inflorescences. All stems, leaves, and calyces are covered with appressed white hairs. Among the hairs, there are scattered glands, yellow-green on the stems, purple on the leaves, and red on the sepals.

The shrubs of this species cover the lower part of Eureka Dunes, where sand mobility is minimal. However, when covered by sand, the plant keeps growing and produces adventitious roots from thick, well-lignified branches. When exposed, the trunk and the thick root may remain alive for a long time. I did not see the plant in sections of the dune with high sand mobility or where only a shallow layer of sand covered the gravel plain.

According to Billings (1945), this plant (recorded as *Dalea polyadenia*) is a dominant shrub of a plant community developed on stable dunes in the Carson Desert, Nevada.

5.2.8 *Sarcobatus vermiculatus*

This shrub exists in many plant communities from Mexico to Canada but is most abundant in cold deserts north of 37° latitude (Billings 1945; Fireman and Hayward 1952; Munz and Keck 1963; Robertson 1983). It is a halophytic phreatophyte (Robinson 1958). The following description is based mainly on my observations of it at sand dunes near Keeler, Owens Valley, and at the old northeastern terraces near Mono Lake, California.

It is a spinescent shrub with succulent, winter-deciduous leaves. In warmer areas, such as the Mojave desert, it can be evergreen. The degree of spinescence seems to be related to local environmental conditions. Plants under water stress become more spiny, as described in the case of *Convolvulus lanatus* (Sect. 5.5.1). It has a high plasticity of reactive growth patterns. It may become dominant in habitats rich in salt (Billings 1945; Fireman and Hayward 1952). Being the most extensive halophyte of the intermountain area (Robertson 1983), it may germinate and establish itself on saline soils on receding lake shores (Flowers 1934). Such soils are also found in large areas at the newly exposed coastal shores of Mono Lake and Owens Lake. I also found seedlings among branches in old nebkas, near Owens Lake. The shrubs induce sedimentation of the sand and finer particles to their leeward, or within their canopy boundaries. When covered by sand, the branches produce adventitious roots. New branches sprout just above these new shootborne roots and a nebka is formed. Taproots have been traced to a depth of 5.7 m (Robertson 1983) and hence, while being covered by sand, the plant keeps drawing water from the deep saline substrate as well as from the sand (JH Richards, pers. comm.).

The young inverse cone-shaped plant (as in Fig. 24c) is at first not an efficient sand trap. However, as it keeps growing above the sand, it can form a phytogenic dune 4–5 m high and dozens of meters long. Prominent longitudinal dunes, covered by nebkas of *S. vermiculatus*, occur on old terraces of Owens Lake 15 km SSE of Lone Pine. The sand trapped by the plant, and its complete cover of the fine-grained older strata, where it started as a pioneer, puzzled Rollins et al. (1968). They stated that, while *Sarcobatus vermiculatus* tolerates high sodic clays and very weak drainage, it thrives in sand. They found that among the largest plants of this species are those on dunes with less than 0.1% salt content in the first 30 cm. The interdune or internebka areas, and the areas below the nebkas in the area of Lake Mono, are ancient lake or lakeshore deposits and include pebbles, pumice, and the remnants of calcitic lake deposits. The *Sarcobatus* nebkas are composed mainly of fine to coarse sand with no remnants of lake deposits above the level of the layer of lake deposits. This particle distribution is a result of the shrub functioning as a trap for airborne sand. Therefore there are hardly any particles larger than sand in the nebkas. I have seen such nebkas in Mono Valley, up to 10 km NNE of the 1941 coastline (1941 was the date of the onset of a substantial artificial drop in the lake level). The sand of the *Sarcobatus* nebkas has a lighter colour than the surrounding ground. This is probably because salts have been accumulated by the recycling of salt-rich leaves deposited on the soil surface (Fireman and Hayward 1952). A high concentration of boron near the *Sarcobatus* shrubs was found to be a factor in preventing colonization by other plants (Rollins et al. 1968). As long as the *Sarcobatus* plants survive, the salt "pump" keeps operating, as found by Fireman and Hayward (1952) for shrubs of increasing age. Meanwhile the internebka soil is leached and, succession leads to the establishment and dominance by *Chrysothamnus nauseosus* in the area close to the present-day lake shore and by *Artemisia tridentata* farther off. Both these species are nonhalophytes and much less resistant to saline or alkaline soils than is *Sarcobatus vermiculatus*. As is true with many other halophytes (Danin 1978a), *S. vermiculatus* will not establish itself in nonsaline soils in the presence of nonhalophytic competitors. The existence of *Sarcobatus* nebkas in a shrub steppe of glycophytes may therefore be due to its ability to change soil salinity locally. The roots seem to have rhizosheaths (J. Richards, pers. comm.), but the sand is not affixed to them as tightly as it is in the grasses. *Sarcobatus* shrubs growing on alluvial terraces of the Mojave River, near Barstow, display obvious rhizosheaths on the thick buried branches of the shrubs forming nebkas. In this site, rhizosheaths were found also on adventitious roots on thick buried branches of *Atriplex polycarpa* and of *Tamarix* cf. *ramosissima*.

A special reactive growth pattern of *Sarcobatus* is the sprouting of rootborne shoots from exposed roots (Robertson 1983). I have seen this property in Owens Lake populations of *Sarcobatus* at the margins of nebkas under deflation. This attribute is the cardinal adaptive character of the *Moltkiopsis ciliata* type (Sect. 5.4). However, unlike species in the *Moltkiopsis*-type syndrome, thick roots of *Sarcobatus* detached from the mother plant cannot produce adventitious shoots. Since *Sarcobatus vermiculatus* does not germinate, establish, and respond to deflation on leached sand, as do the plants discussed in Section 5.4, and does not reproduce from thick detached roots, it is placed in the *Calligonum comosum* group.

5.2.9 Summary

Plants of this group typify the second step of plant succession in desert dunes. They are capable of withstanding covering by sand and of sand deflation (with consequences of root exposure), and of growing on stable sand. Some of them may have deep tap roots several meters long and use water resources below the sand layer. The common attributes of the species of the *Calligonum comosum* type syndrome are:

1. All the dicot species of this group have the ability to produce adventitious roots from old sand-covered lignified branches.
2. New branches sprout above the locations where new adventitious roots are developed.
3. Many of the species listed above, even the grasses, are stem-assimilants.
4. The roots of the dicot shrubs of this group can withstand long exposure; the grasses and at least one of the dicots have rhizosheaths which may protect the exposed roots.
5. The grasses in this group are branched, have rhizodes, and may expand in three different ways, depending on sand mobility.

5.3 Species of Areas with Modest Sand Cover or Removal–*Stipagrostis plumosa* Type

The plants of this group grow mostly in sites where there is not much sand accretion or deflation. However, they may withstand such hazards by the elongation of buried stems and the production of shootborne roots. They differ from the plants in Section 5.2 in the smaller amounts of sand involved.

5.3.1 *Stipagrostis (= Aristida) plumosa*

This is an amphiphytic perennial or rarely annual grass, 15–50 cm tall, that may grow on stable or modestly mobile sand. The basal parts of the shoots constitute the brachyblastic sympodial leafy rhizome (Fig. 31a–c). The basal 1 cm of each stem has a few short internodes, the lowest of which have no axillary buds, whereas the upper ones have buds protected by leaf bases and dry scales. The lowest 2–3 aerial internodes of each shoot are tomentose and appear whitish green. There is a bud in the axil of each leaf of the aerial parts of these shoots. Intravaginal tillering is the rule in *S. plumosa*. When moisture is adequate the shoot may have green leaves, terminate in an inflorescence and have tillers (Fig. 32). When moisture is limiting, the plants will not bloom and fail to produce new tillers. They decrease the sizes of the active crowns to the lower 2–3 green and tomentose internodes of a few tillers. When moisture returns, new shoots sprout as a continuation of the leafy rhizome at the base of the vertical shoot (Fig. 31b–c). When aerial shoots are covered by small quantities of sand (A in Fig. 31d), new nodal roots, with rhizosheaths, may be produced. The lateral bud which was buried and the one above it produce tillers with further lateral buds which terminate in an inflorescence.

It is rare to find this species in places with high sand mobility. It is found in places where the sand does not move (Fig. 33; Sects. 6.1 and 6.2), and also in deflation sites and where it survives even when the roots and rhizosheaths are

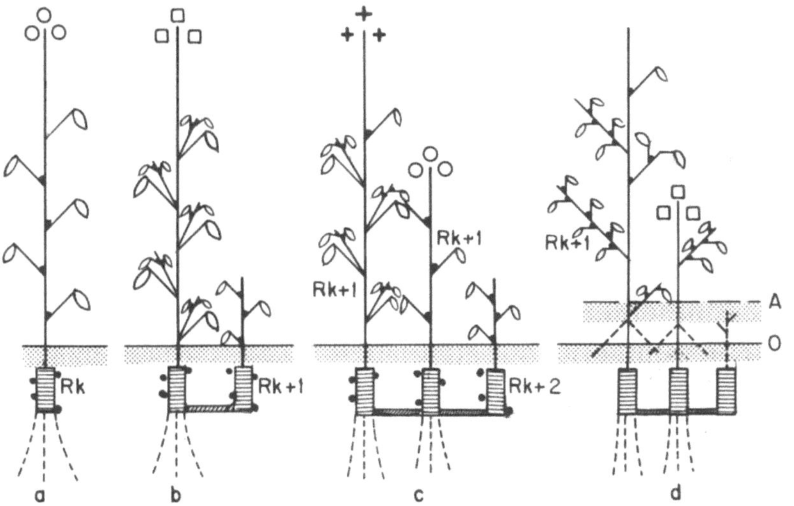

Fig. 31a–d. Schematic presentation of structure and development of *Stipagrostis plumosa*: **a** A flowering ramet of the R_k order with a bud at the axil of all leaves; **b** R_k ramet with a few branches and a new ramet (R_{k+1}) developed from a bud on the subterranean brachyblastic part of R_k; **c** as b with a new ramet (R_{k+2}) **d** response to cover to level *A* – development of nodal roots at the proper level

Fig. 32. One ramet of *Stipagrostis plumosa* whose central stem terminates in an inflorescence (out of the photo); at the base, two lateral branches connected by a minute leafy rhizome are starting to develop. One root with rhizosheath extends from the lower internode

Fig. 33. A tuft of *Stipagrostis plumosa* on a stable sand dune, near Revivim, in the northern Negev

Fig. 34. The uprooted tuft of Fig. 33 displaying rhizosheaths on all roots. The ramets are held together by many very short leafy rhizomes

exposed (Fig. 34). It may be that the relatively dense horizontal structure of the plant (Fig. 33) prevents sand accumulation within the tuft. A vertical cross section through the plant has the shape of an inverted pyramid; perhaps this shape is an inefficient windbreak.

The diaspore (Fig. 23b) is of the "gliding type" (Sect. 5.1.1). The central awn is plumose, straight, and longer than the lateral glabrous and straight awns. The awn's base is slightly twisted, rotated 1–2 times, and probably functions as a rudimentary drill. The seminal root (developing from the embryonic root) is poor in root hairs. Only rarely do seminal roots have rhizosheaths and those are limited to the tip. The seedling's nodal roots are, however, densely covered with root hairs, while a thick rhizosheath extends from the proximal to the distal part of the root.

5.3.2 *Stipagrostis* (= *Aristida*) *drarii*

This is an amphiphytic perennial grass, 50–100 cm tall, capable of growing on stable or modestly mobile sand. The ramet is made up of a stem which is the distal section of a short plagiotropic leafy rhizome. Stems have short (2–3 mm long) internodes at their leafy basal part and terminate in an inflorescence. The inflorescence portion of the stem has 3–4 internodes, each 2–4 cm long. The leaf blades are convolute throughout, horseshoe-shaped in transverse section, and have buds at the axil of most leaves (L_2–L_{11}). Tillers develop in an intravaginal pattern once their main apex becomes an inflorescence.

This species may be regarded as amphiphytic because of the two possible modes of growth. If not covered by sand, the plant produces branches like the ramet mentioned above, starting from the brachyblastic section. It can survive burial of up to 25 cm from the branched plant. In time, buried nodes produce new roots and tillers and much of the circular tuft starts growing again. It has the potential of growing even if covered by 25 cm of sand, but is typically found on shallow sands along with *S. plumosa*. Its awned dispersal units resemble those of *S. plumosa* (Fig. 23b).

5.3.3 *Centropodia (= Danthonia, Astenatherum) forskalii*

This is an amphiphytic perennial or rarely annual grass 5–70 cm tall, of desert and Mediterranean areas. It is capable of growth in stable, slightly mobile, or mobile sand. The annual form will be discussed in Section 5.8.2.

The ramet is the terminal living part of the leafy rhizome with vertical aerial shoots. The basal brachyblastic parts of the shoots form the rhizome. The roots emerge from nodes buried in the sand. The basal 1 cm of each stem has a few short internodes, the lower ones have no axillary buds, whereas the upper ones have buds protected by leaf bases and dry hairy scales. The basal brachyblastic section is thicker than the aerial dolychoblastic part of the shoot. The aerial internodes of the culm are covered by appressed hairs, most of which are basipetally directed. The leaf sheaths are also covered by similar hairs, situated mainly in longitudinal depressions among the veins. There are axillary buds throughout these stems, each having a hard outer, hairy scale. The sprouting of these buds depends on the nature of the sand cover. If it is minimal, the resulting leafy rhizome will be thick; if there is more sand, the rhizome will include sections of aerial stems that are thinner than the brachyblastic part of the ramet. Distal tillers on aerial shoots far from the roots die. The result of this plasticity is that every individual, as a result of its own local history, looks different.

When moisture is adequate shoots have green leaves, tillers, and terminate in an inflorescence. The length of stems and number of inflorescences depend largely on the quantity of water available during the growth season. When moisture is limited, the plant may not bloom and will fail to produce tillers. The dry leaf sheath remains on the plant for more than 2 years, and the basal internodes of the aerial shoots may stay green for an equally long time. The distal roots emerge from plagiotropic leafy rhizome sections 1–3 years old. The life cycle resembles that of *Stipagrostis plumosa* (Fig. 31). *Centropodia forskalii* has a hairy diaspore with one caryopsis subtended by a palea and lemma. The lemma is hairy with short appressed hairs on the lower part and long spreading hairs on the upper. These may contribute to wind dispersal either by rolling, as in *Calligonum* species, or by creeping as in *Stipagrostis scoparia;* gliding is not likely to occur.

In some habitats this species may fit better the groups in Sections 5.1 or 5.2. However, in the majority of sites it belongs to the habitat and plants in Section 5.3.

5.3.4 *Cyperus macrorrhizus* (= *C. conglomeratus*)

For a long time this plant was known in the literature as *C. conglomeratus* (Eig 1939; Feinbrun-Dothan 1986; Danin and Kukkonen 1995). It is a geophyte confined to mobile or semi-stable sands in the desert and Mediterranean parts of Israel (Eig 1939).

The ramet is a tuberous rhizome, composed of 4–5 internodes with a viable dormant bud at the axil of each leaf (Fig. 35a). The outer scales of the buds are hard and dry. The plant is sclerophyllous; its leaves and the scape of the inflorescence have a thick, shiny cuticle. The leaves have a lamina 3–4 mm wide that folds adaxially to form a pipe of 1–2 mm in diameter. A leaf sheath covers the stem or the younger leaves. The sympodial stem terminates its growth by blooming and fruiting (Fig. 35a,b).

A dry remnant of the scape terminates the tuberous sympodial axis. The present year's leaves are attached just below the inflorescence (Figs. 35b, c). The first internodes of the tiller may be a few millimeters or a few centimeters long - depending on the amount of sand covering the plant (Figs. 35c, 36b,c, 37). During the growing and blooming seasons, the tuber's first few internodes swell and expand into their final size, while the terminal bud becomes an inflorescence. A bud, just below the inflorescence culm, is the renewal bud which will become next year's stem terminating in an inflorescence. Its base is the extension of the rhizome.

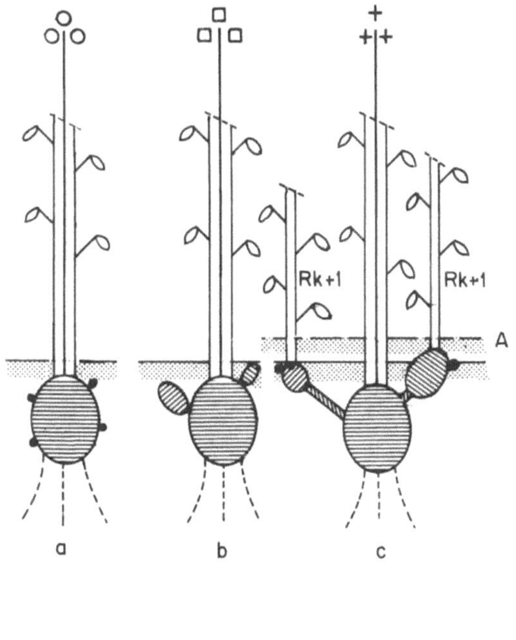

Fig. 35a–c. Schematic presentation of the structure and development of *Cyperus macrorrhizus*. **a** A flowering ramet with tuberous brachyblastic base; **b** bud starts to develop after fruiting; **c** growth of ramet R_{k+1} with an elongated internode as a response to sand cover (to level *A*)

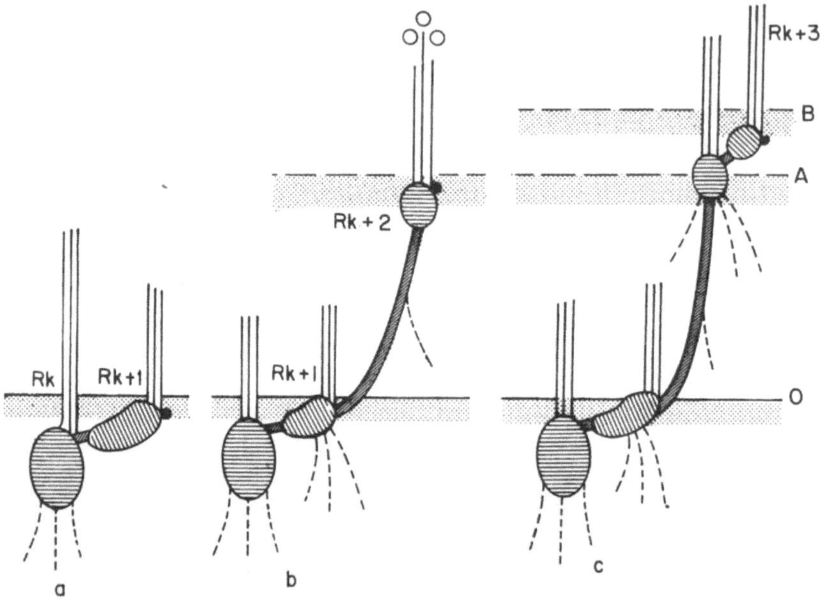

Fig. 36a–c. Response of *Cyperus macrorrhizus* to sand cover. **a** Regular growth of ramet R_k at level *0* with extension of the new ramet R_{k+1} (leaves not drawn); **b** response of the plant to sand cover to level *A* by elongation of the first two internodes of the brachyblastic part of the ramet R_{k+2} and formation of tuber and inflorescence; **c** response to additional cover to level *B* by a change in the direction of growth of the brachyblastic base of R_{K+3}

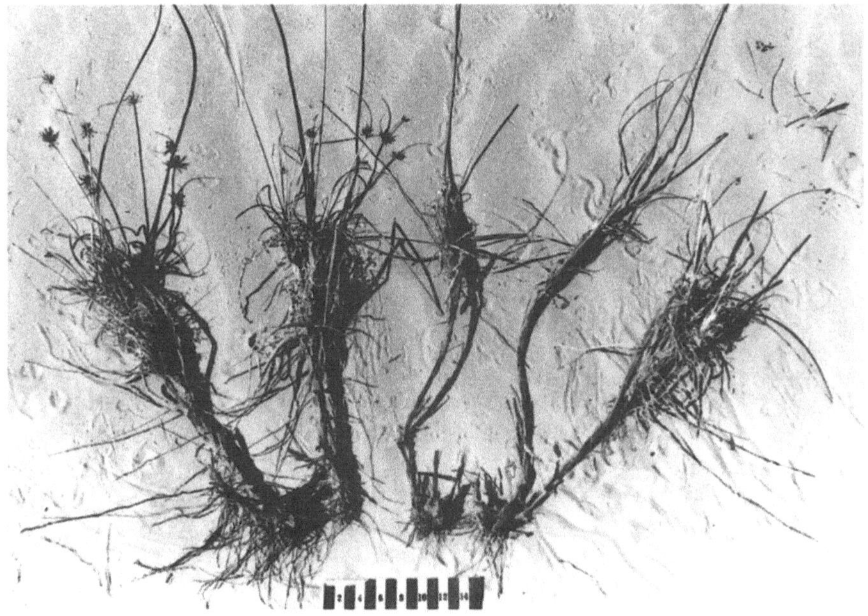

Fig. 37. Response of *Cyperus macrorrhizus* to sand cover (for explanations see Fig. 36)

The tuberous part of the rhizome remains alive for at least 8 years, even when deeply covered by sand, and most of the buds remain dormant with dry brown scales and leaf bases covering the viable yellowish-green soft leaf primordia. Gimingham (1964) found living buds of *Elymus farctus* at a depth of 60 cm in coastal dunes of Scotland. The reservoir of buried buds below the dune surface is a "bud bank", as used by Harper (1977) in a somewhat different context. Part of the bud bank becomes active once tubers are exposed by sand deflation (Fig. 38).

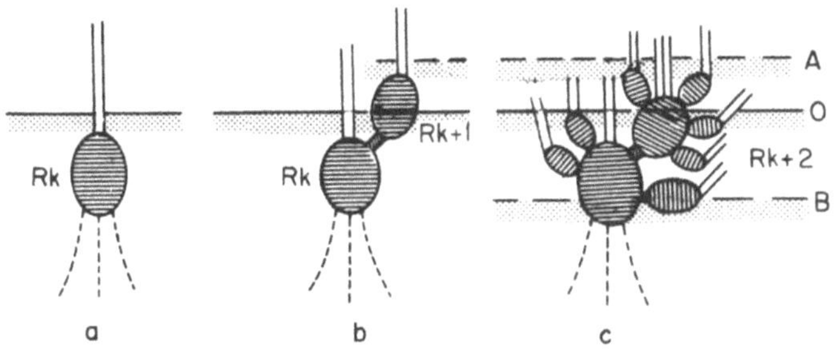

Fig. 38a–c. Response of *Cyperus macrorrhizus* to removal of sand. **a** Ramet R_k growing at level *0*; **b** growth of R_{k+1} as a response to cover to level *A*; **c** sprouting of several R_{k+2} branches as a result of exposure of tubers to level *B*

Fig. 39. Rhizosheaths on a root system of *Cyperus macrorrhizus* from the Negev. The fully active roots on the right ramet have thick rhizosheaths

Roots with well-developed rhizosheaths (Figs. 39–42) come out of the tuberous part of the rhizome and burst through the bases of the previous year's dry leaves. The roots may function for 1–2 years. The ability of the plant to develop rhizosheaths increases its resistance to root exposure. The functions of rhizosheaths are further discussed in Section 7.3.4.4.2.

The mobility of the plant, expressed by its reactive growth, demonstrate high plasticity concerning growth rate and the direction of expansion of the rhizomes. When the sand is relatively immobile, the plant produces leafy rhizomes which grow from lateral buds on the previous year's vertical stem (Fig. 35b). The old ramets of the plant, situated at the center of the tuft, die and deteriorate while the young ones keep growing at the periphery. As a result, an old tuft on stabilized sand may become a circle of living ramets 50–100 cm in diameter (Fig. 43). Being protected from wind erosion, the center may fill up with sand and support annuals. I saw such a pattern in the Mediterranean section of its distribution area in Israel, 10 km SE of Netanya. The specimen in

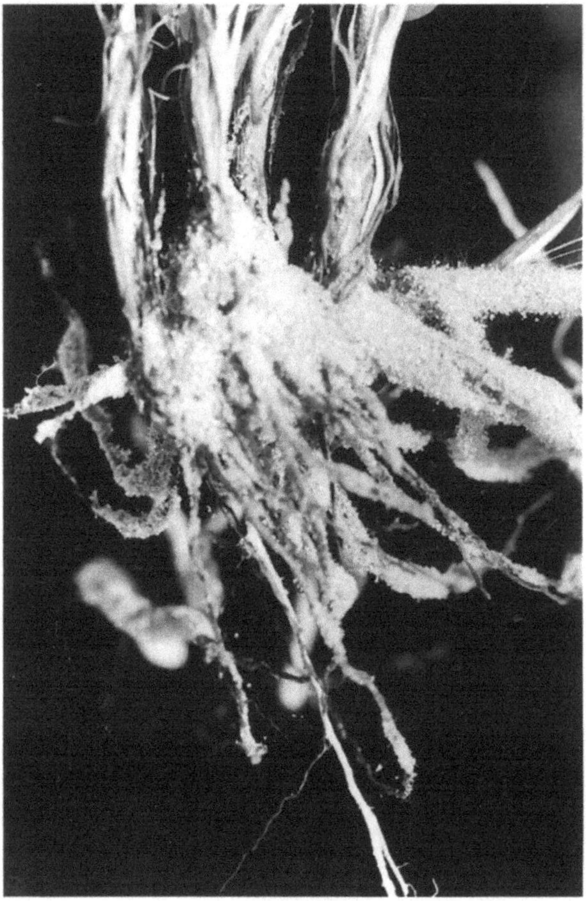

Fig. 40. A close-up of roots with rhizosheaths; note the nearly free root hairs at the periphery of the horizontal uppermost root on the *right*

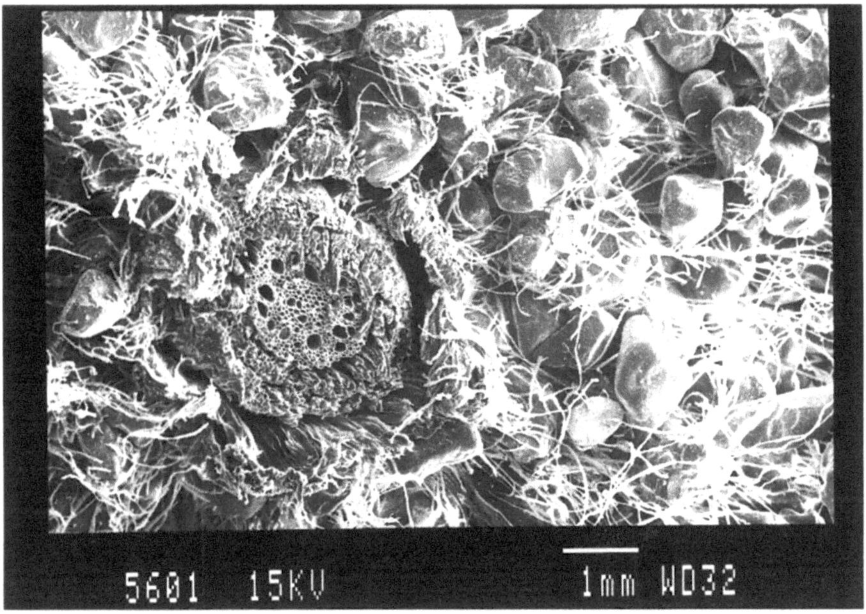

Fig. 41. SEM micrograph of a rhizosheath of *Cyperus macrorrhizus* with root hairs visible among the sand grains

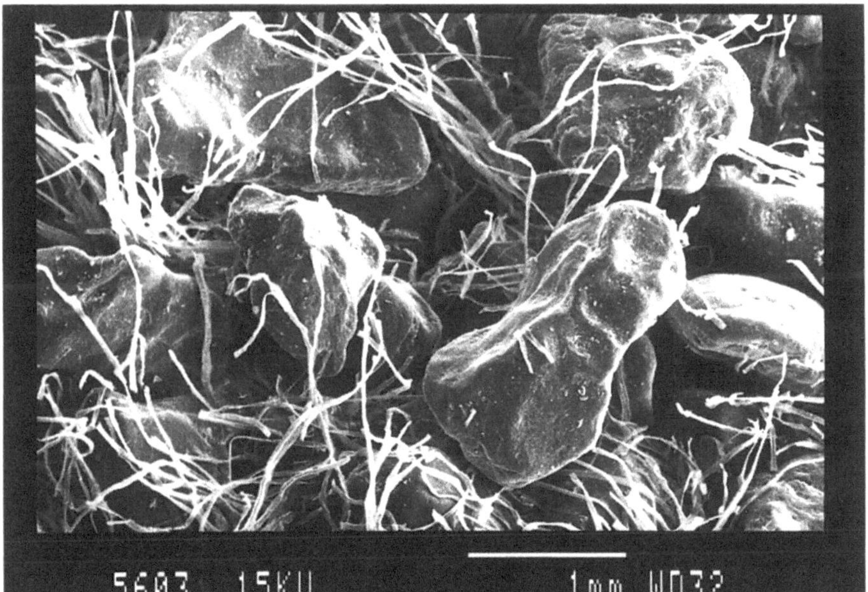

Fig. 42. SEM micrograph of a few sand grains in a rhizosheath; note remnants of root hairs

Fig. 43. A circle of *C. macrorrhizus* on a sand sheet, northern Negev

Fig. 37 displays at its lowest part a plagiotropic rhizome which developed during years when no accretion or deflation of sand took place. The addition of a few centimeters of sand in recent years killed the leaves but did not harm the buds on the rhizome. Sand cover induced the development of a vertical, negative geotropic section in the rhizome (as discussed for Fig. 36). When growing vertically, the old parts remain buried in the ground and the tuft above the ground creates a small round nebka, but not a circle of independent ramets.

When covered by sand, the plant produces vertical tillers in an intravaginal pattern above the tuber. The base of each tiller is composed of 1–2 internodes (Figs. 36b, 37, 39) longer than those on stems which have not developed under sand cover (Fig. 35c). Whereas the tubers of plants in stable sand have internodes 2–3 mm long, the specimen in Fig. 37 had internodes 10–30 mm long. The elongation of internodes can take place even with the internodes of the tuber. Consequently, a vertical stem there develops with 2–3 small tubers made up of 1–3 swollen internodes each. The modes of reactive growth dealt with in this species attest the occurrence of intercalary meristems along the geocorms. The activity of such meristems in the internode growth of monocots is discussed by Fahn (1982). A plant growing in a site with steady accretion forms a small nebka without a dead center because tillers grow in all directions, not just towards the periphery.

The sequence of events which brought the specimen in Fig. 38c to its present shape was as follows. After germination and establishment (Fig. 38a,b) the plant was covered by a few centimeters of sand (line A) and produced the branch R_{k+1}. As a result of exposure of the entire rhizomatous section of the plant by subsequent wind erosion (line B in Fig. 38c), the latent buds situated on the previous years' tubers produced a leafy tiller (R_{k+2}). Buds buried for a few years at a depth of 20 cm are still capable of sprouting.

When movement of the sand is arrested, *C. macrorrhizus* continues growth to a limited extent until eventually outcompeted by plants better adapted to stable sand. In an area 10 km SE of Netanya, where sand was mobile for many years and became stable only some 15–20 years ago, most of the tufts of *C. macrorrhizus* are now dead. Each tuft is connected to a horizontal leafy rhizome with 15–20 tubers. The reactive growth of this species and its tuft morphology are environmental indicators. Ring-shaped tufts indicate stable sand, whereas nebkas with tillers in all directions indicate sand accretion. Preservation of rhizomes in dry sand may even permit palaeoecological deductions.

The dispersal unit is a nutlet 2–2.5 mm long.

5.3.5 *Cyperus conglomeratus*

This species is closely related to *C. macrorrhizus* in its morphology but differs in habitat and fruit micromorphology (Danin and Kukkonen 1995). The plants examined were collected in S Sinai, on an ancient fossil reef near Ras Nusrani where they grow on sandy ground. By trapping airborne grains they enhance sand accumulation above the fossil reef (Danin 1983). The stems have basal leafy brachyblastic tuberous rhizomes composed of 7–10 short internodes. The roots, covered with rhizosheaths, emerge from rhizome nodes, mainly from the tuberous section when it is 1–3 years old.

5.3.6 *Echinops philistaeus*

This is an evergreen chamaephyte, 40–60 cm high and capable of surviving being covered by small quantities of sand. It grows in Israel and N Sinai under Mediterranean conditions (400–500 mm mean annual rainfall) and under desert conditions (80–100 mm). When covered by sand, it is capable of producing shootborne roots. The roots have no rhizosheaths. A specimen collected near Ashalim, N Negev (Shunra-Nitzzana sands) had lignified stems with ten annual rings and thick bark.

5.3.7 *Tiquilia* (= *Coldenia*) *plicata*

This is a perennial herbaceous plant of southwestern USA (Kelley in Hickman 1993). It was the dominant herbaceous species among the nebkas of the lower stable section at Eureka Dunes (Sect. 6.7.3) when I visited them in September

1994. The only plant parts seen above the ground were small leaf rosettes. The leaf surface is densely covered with appressed hairs. The lower side of the leaf, leaf petiole, and aboveground parts of the stem have many glandular trichomes which secrete a sticky substance. Such trichomes are known as colleters (Fahn 1982). Sand particles become attached to the viscid material. The structure of plant parts below ground is responsive to sand accretion. Young individuals which have not been covered by much sand, have roots beginning a few centimeters below the sand surface. The distal parts of some of the roots become thick, funnel-shaped root tubers, rich in parenchymatous storage tissue. When covered by a thin layer of sand, new leaves and thin stems 5–20 cm long emerge above the root. Buried stems have affixed sand, depending on their position and history. Stem parts that have been exposed to strong abrasive winds are devoid of sand cover, whereas parts which have not been exposed to strong winds later became covered by sand and may still have a layer of sand affixed to them. Buried stem nodes keep their sand cover for a longer time than do the internodes because there are both simple hairs and colleters at the nodes. Herbarium specimens show that plants not growing in sites with sand accretion, have lignified tap roots capable of giving rise to new prostrate stems. Hence, the individual structure of each specimen represents its reactive growth in the local sand cover at its site. Long vertical stems below ground may give rise to shootborne roots which tap water from a shallower depth than that of the old buried root system.

Young roots have rhizosheaths, a phenomenon rare among dicotyledons (Sect. 7.3.4.4.2). Remnants of persistent root hairs are seen even in old herbarium specimens. *Tiquilia* (= *Coldenia*) *nuttalii* when growing in sites with mobile sand, is a prostrate annual with sand-covered leaves, petioles, and young stems.

5.3.8 Summary

The seven representatives of this ecomorphological type grow in desert dunes that are nearly stable. They are capable of withstanding modest sand accretion or deflation. They also have enormous ecomorphological plasticity. The grasses among them share a few additional properties.

1. Plasticity among grasses is primarily displayed by their ability to shift from being annual to perennial according to the amount of available water.
2. The grasses have axillary buds along the culm, enabling them to escape being covered by sand accretion. *Echinops philistaeus* is the same.
3. All the monocots (and one dicot) have rhizosheaths which may function in the improvement of moisture and nutrition resources.
4. All the species have a protection against airborne sand abrasion: the dense woolly cover of the grass culms, stems and leaves of *Echinops*, the affixed sand of *Tiquilia*, and the sclerophylly of the *Cyperus* species.

5.4 Species Actively Resistant to Sand Deflation – *Moltkiopsis ciliata* Type

The plants belonging to this ecomorphological group have a reactive response to sand deflation. When their roots are exposed, they produce adventitious shoots from the exposed roots. All are Boraginaceae and occur in the Old World.

5.4.1 *Moltkiopsis ciliata* (= *Lithospermum callosum*)

This is a semishrub confined to semistable sand or sites of sand deflation. Seedlings are to be found in sites of modest deflation and in large blowouts. If sand is relatively stable, the plants keep growing. They produce long tap roots, at least 1–2 m, with lateral roots that branch at right angle to the taproot (Figs. 44, 45c). The lateral plagiotropic roots run 2–3 m parallel to the soil surface and then become geotropic (Figs. 44, 45b). When exposed, they produce rootborne vertical shoots (R_{k+1} in Fig. 44c). Even when covered with sand, these adventitious shoots do not produce adventitious roots. Lateral roots, exposed here and there, may survive for a few years after the mother plant has died (Fig. 46). The roots of this species are succulent and have a storage layer in the secondary phloem and cortex. The storage material stains pink with Sudan IV, indicating lipid. The leaves are covered with appressed white hairs and spreading bristles at the leaf margins.

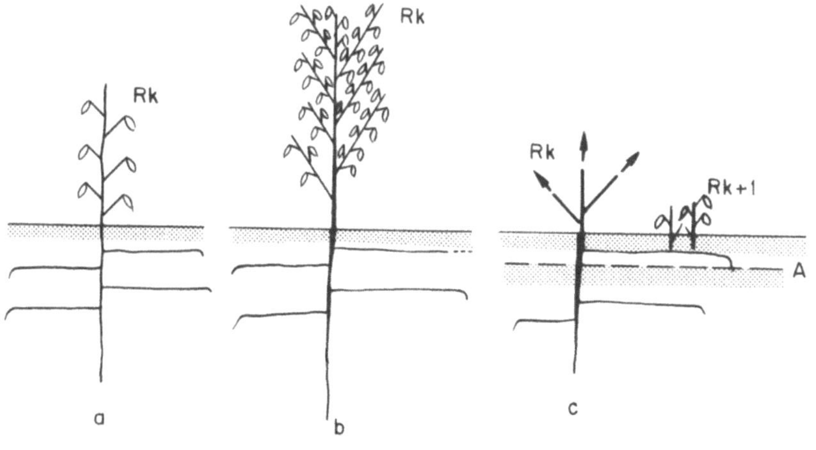

Fig. 44a–c. Schematic presentation of structure and development of *Moltkiopsis ciliata*. **a** and **b** Growth and development of bimodal root system; **c** sprouting of adventitious shoots from roots after being exposed to level *A*

a

b

Fig. 45a–c. Root system of *Moltkiopsis ciliata.* **a** With complete sand cover; **b** after removal of sand; **c** the exposed vertical tap root and horizontal lateral roots

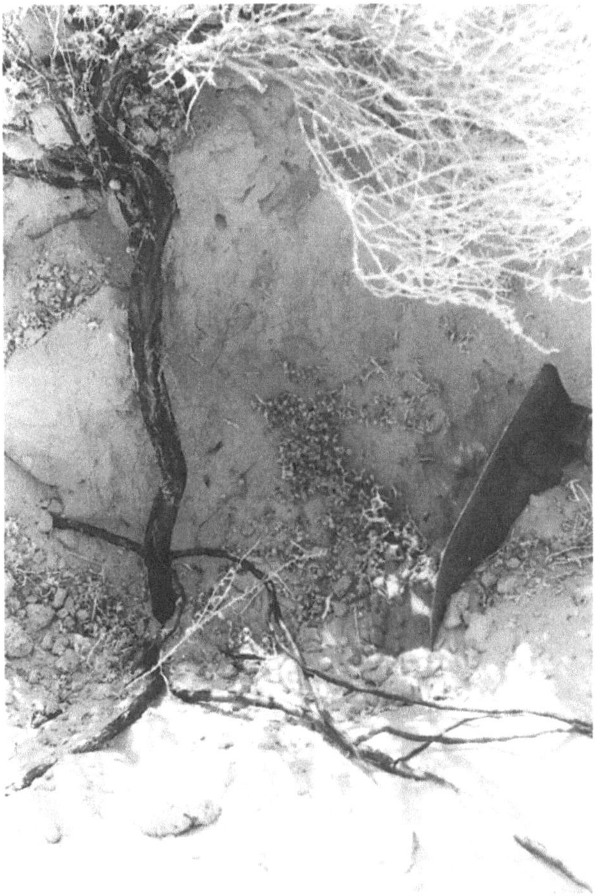

Fig. 45c

c

5.4.2 *Heliotropium digynum* (= *H. luteum*)

This species resembles *Moltkiopsis* ciliata but is restricted to desert sands. In many places it is deciduous in summer, whereas the two species mentioned above are evergreen. The succulent roots of this species (Fig. 47) have starch grains in the parenchyma of the secondary phloem and probably in the cortex as well. *Heliotropium arguzioides* of Central Asia (Walter and Box 1983a) appears to be the same.

5.4.3 *Echiochilon fruticosum*

This is a common semishrub confined to blowouts in sand sheets or to stabilized dunes in the Israeli desert and Mediterranean coastal sands. Its root system is lignified and not succulent, unlike those of *Moltkiopsis ciliata* or

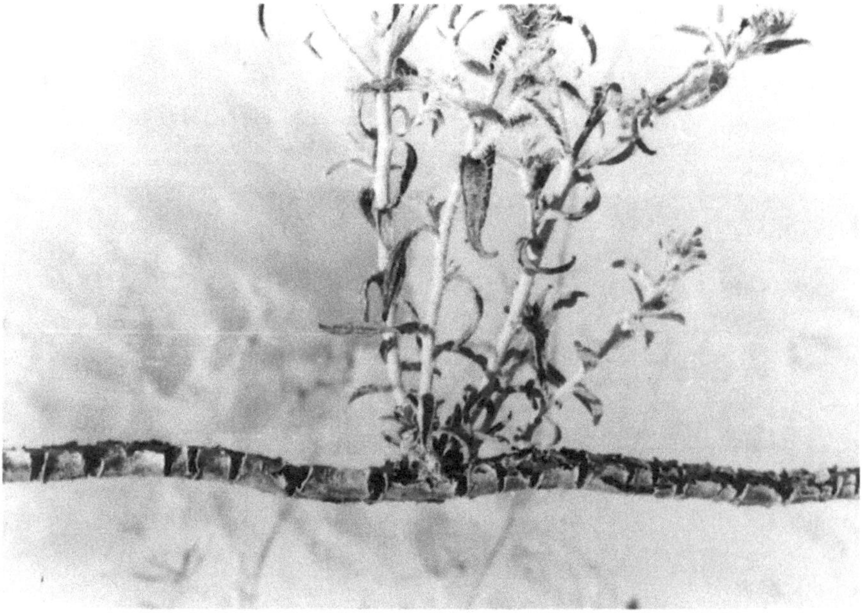

Fig. 46. Rootborne shoots of *M. ciliata*

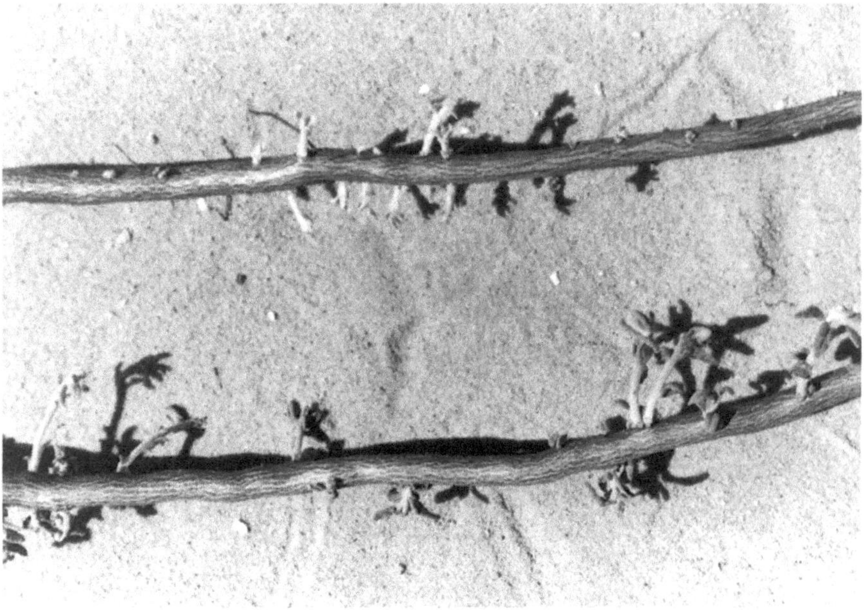

Fig. 47. Rootborne shoots of *Heliotropium digynum*

Heliotropium digynum. The leaves are densely covered with appressed white hairs and the whole plant looks grey.

5.4.4 Anchusa negevensis

This perennial herbaceous plant was discovered when material for this book was collected in the Negev (Fig. 48; Danin 1995b). It is endemic to an area of

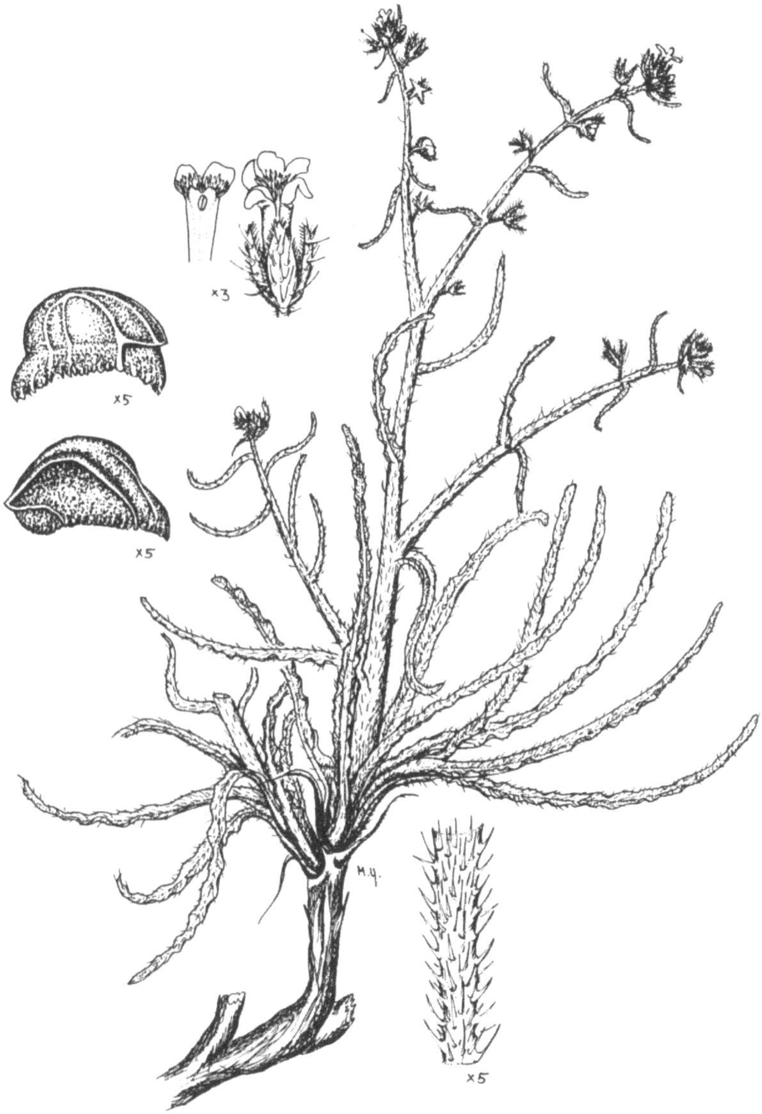

Fig. 48. *Anchusa negevensis,* a desert plant adapted to sand deflation by its ability to produce adventitious shoots from succulent roots

about 0.2 ha on a slope formed by an outcrop of weathered Neogene sandstone, where it is exposed to high wind erosion. It has succulent roots capable of producing rootborne shoots, and illustrates the theme of this book. We expect to find the same ecomorphological type of plant repeatedly in similar habitats regardless of the floristic region because morphological types such as this represent convergent adaptations. The specific slope on which this plant was found, was dominated by plants adapted to sand deflation. The ecomorphological type and the adaptations of *Anchusa negevensis* had already been predicted well before it was collected, named, and formally described.

5.4.5 Other Species with Rootborne Shoots

Several species with the ability to produce adventitious shoots grow in sandy areas but do not posses the syndrome discussed in Sections 5.4.1 through 5.4.4. *Lycium schweinfurthii* is a phanerophyte and a component of climax or near-climax communities on sand sheets in desert and coastal areas of Israel and Sinai. It has small black berries which are dispersed by birds. Most individuals in one patch, of the study area of Section 6.1, a few tens to hundreds of meters square, arose as adventitious shoots from lignified roots. The plant propagates in this way with no evident relation to sand deflation. I have never seen seedlings of this plant, and germination and establishment must be rare event. *Prosopis farcta* is a phreatophyte which occurs in the Negev where sand has recently covered the original habitat (Sect. 6.1). It produces adventitious shoots in all the other habitats throughout Israel in which it grows (Dafni 1975), and cannot be regarded as being adapted to sand in any way, although it may grow in sandy areas. In the same way, *Acacia albida* is found in the coastal dunes of Israel near Ashdod, and may occur in desert sandy areas throughout its large area of distribution. It reproduces from adventitious shoots all over Israel, even when growing as a phreatophyte in the Bet-Shean Valley. *Sarcobatus vermiculatus* is one of the few plants found in sand in N America which has the ability to reproduce from exposed roots. However, as discussed in Section 5.2.8, it is basically a phreatophytic halophyte. Even when growing on sand, it is more like the *Prosopis farcta* and *Acacia albida* than *Moltkiopsis ciliata* type.

5.4.6 Summary

Plants of this group are psammophytes which do not grow in nonsandy habitats. They may grow for several years as regular components of stable sand communities. When deflation takes place, as a result of local blowouts or other reasons, they are able to produce rootborne new shoots. Even when the mother plant dies, its roots may remain alive for some time and produce new shoots.

5.5 Passive Resistant Species to Sand Deflation – *Convolvulus lanatus* Type

Plants of this group have hardy tap roots with thick corky bark or other means of protection from the exposure consequent on deflation. They are found either in areas of constant deflation or on stable sand sheets. When covered by sand they do not produce shootborne roots.

5.5.1 *Convolvulus lanatus*

This is a grey semishrub densely covered with spreading hairs. When exposed, the root displays thick, cracked bark (Fig. 49). The plant keeps on growing and carries out all its functions, such as blooming and bearing fruits, for many years. The stems do not generate shootborne roots when they are covered by sand. The species dominates in sandy areas which are under constant deflation. In northwestern Sinai there are considerable areas of old, stable dunes (area 2 in Fig. 3) with a few scattered mobile ones (areas 3 and 4 in Fig. 3). *Convolvulus lanatus* dominates these dunes which are constantly losing fine-grained particles by wind erosion. In the western Negev it dominates areas of

Fig. 49. An exposed root of *Convolvulus lanatus* resulting from deflation by wind erosion

constant deflation such as the west-facing slopes of ancient dunes cut by recent wadis. It also grows as a component of various associations on stable sandy habitats. Some of these associations may have had a history of wind erosion during their development. It is also found in the southern coastal areas of Israel and N Sinai. It is a spinescent plant under extreme desert conditions but, in coastal areas and in the northern Negev, it has relatively soft stems. Spines are probably the result of dry periods when stem apices cease growing and lignification occurs (Danin 1983). Under coastal conditions and in the northern Negev, drought is not so extreme and the plant keeps growing without lignification of the narrow region of xylem at the apex.

5.5.2 *Cornulaca monacantha*

This is a spiny semishrub which may attain dominance in locations of stable, sandy and silty soil (such as is found near the Israeli-Egyptian border terminal near Nizzana) and in areas under constant deflation. In areas of deflation, its long tap roots are frequently exposed (Danin 1983). The conductive system of the plant is protected by the special anatomy characteristic of the Chenopodiaceae (Fahn 1982; Fahn and Cutler 1992).

5.5.3 *Retama raetam*

This is a very common stem-assimilant (Lyshede 1977) shrub of many desert and nondesert habitats. In sandy areas with deflation, spectacular shrubs of *Retama* with exposed taproots more than 2 m long (Fig. 50) can be seen. In coastal areas, its roots reach a depth of 20 m (Zohary and Fahn 1952). When covered by sand, the plant often produces a nebka (Fig. 51), but no adventitious roots are found. The vertical roots have xylem vessels with a narrow diameter whereas the vessels of the horizontal roots have a wider diameter (Fahn and Cutler 1992). This appears to be an adaptation of the root system for the efficient conduction of water from the surface layers of the soil which are often wetted to a limited depth.

The diaspore is the entire fruit, which remains closed when ripe and contains 1–3 seeds. It is dispersed by rolling on the sand, by animals eating the fruits and excreting the seeds, and by flood water in wadis (Danin 1983). This species is one of the most common dominants of sandy areas in Israel, and is used in phytosociological syntaxonomy to name its class of Israeli sand vegetation (Zohary 1973). The dominance of *R. raetam* on limestone hills covered by sand is discussed in Section 6.6.

5.5.4 *Thymelaea hirsuta*

This is a very common evergreen microphyllous shrub of many desert and nondesert habitats, including deflation areas. *T. hirsuta* plants survive root

Fig. 50. An exposed root, more than 2 m long, of *Retama raetam* resulting from deflation by wind erosion

Fig. 51. A shrub of *Retama raetam* with a nebka in a stable dune, 10 km S of Revivim, the northern Negev

Fig. 52. An exposed root, more than 1 m long of *Thymelaea hirsuta* resulting from deflation by wind erosion on a slope near Nizzana, the Negev

exposure and keep growing as if they were not exposed at all (Fig. 52). Shootborne roots have not been found.

5.5.5 Summary

The plants listed in this group share only one ecomorphological trait – they all have long taproots which may withstand prolonged exposure. They do not display any prominent reactive growth (other than the production of thick, corky bark).

5.6 Herbaceous Perennial Species of Stable Sand Sheets – *Stipagrostis obtusa* Type

Plants of this group develop in desert sands with low mobility; they rarely develop in sites of sand accumulation. The species in this group belong to the Gramineae and Liliaceae.

5.6.1 *Stipagrostis (= Aristida) obtusa*

This is a short hemicryptophyte or geophyte of sandy desert habitats. The ramet is the distal section of a plagiotropic, brachyblastic, leafy rhizome with

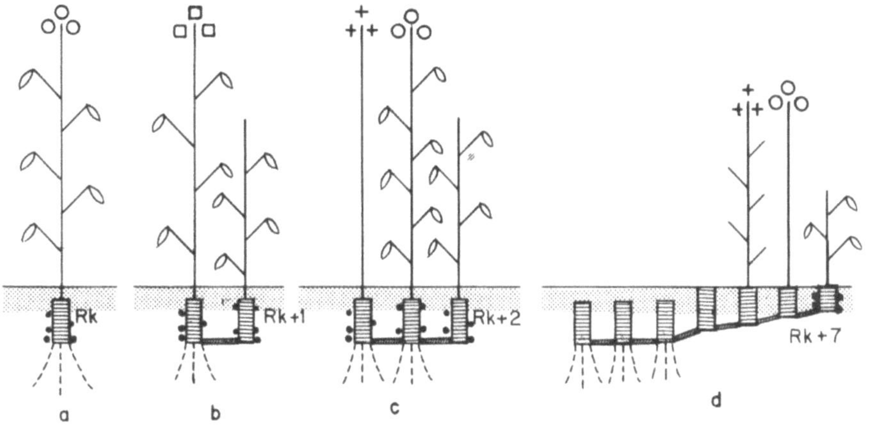

Fig. 53a–d. Schematic presentation of structure and development of *Stipagrostis obtusa*. **a** R_k ramet with basal brachyblastic bud-carrying part, and an inflorescence culm with no buds; **b** R_k ramet and its R_{k+1} branch; **c** R_k ramet, its R_{k+1} and R_{k+2} branches; **d** R_k to R_{k+7} ramets, only those of the last three orders are still alive

stems that have short (2–3 mm) internodes at their leafy basal parts. The stems may terminate in an inflorescence of which the scape has two internodes 5–6 cm long. The leaves are rolled so that their blades are filiform. The number and size of axillary buds decrease from the rhizome towards the inflorescence. There are no buds in the leaf axils of the aboveground culm (Fig. 53a). The number of buds sprouting every year from below each inflorescence is relatively high, giving rise to 4–5 tillers on stems only 10 mm long (Fig. 54b). The old stems die in a centripetal direction (Fig. 53d) and new tillers grow at the periphery. Since buds sprout in all directions, the tuft remains compact or ring-like (Fig. 56).

A few tufts that I studied exhibited 5 cm vertical development of leafy rhizomes. This implies that the plants had been covered by sand and had the capacity for vertical growth of the leafy rhizome. Many tufts had exposed roots, indicating wind erosion. Owing to its dense structure the plant functions as a compact obstacle that leads to wind erosion near the tuft (as shown in Figs. 6–8) rather than to sand accumulation, especially in regions where the wind direction changes frequently.

5.6.2 *Stipagrostis (= Aristida) ciliata*

This is a densely tufted grass which is generally a perennial but, in extreme desert areas such as Namibia and the southern Negev, it may complete its life cycle within a year. It is practically absent from sites of sand accumulation. Its morphology has been studied by Danin and Orshan (1995). The ramet is made up of a stem, which is the distal section of a plagiotropic, leafy rhizome (Fig.

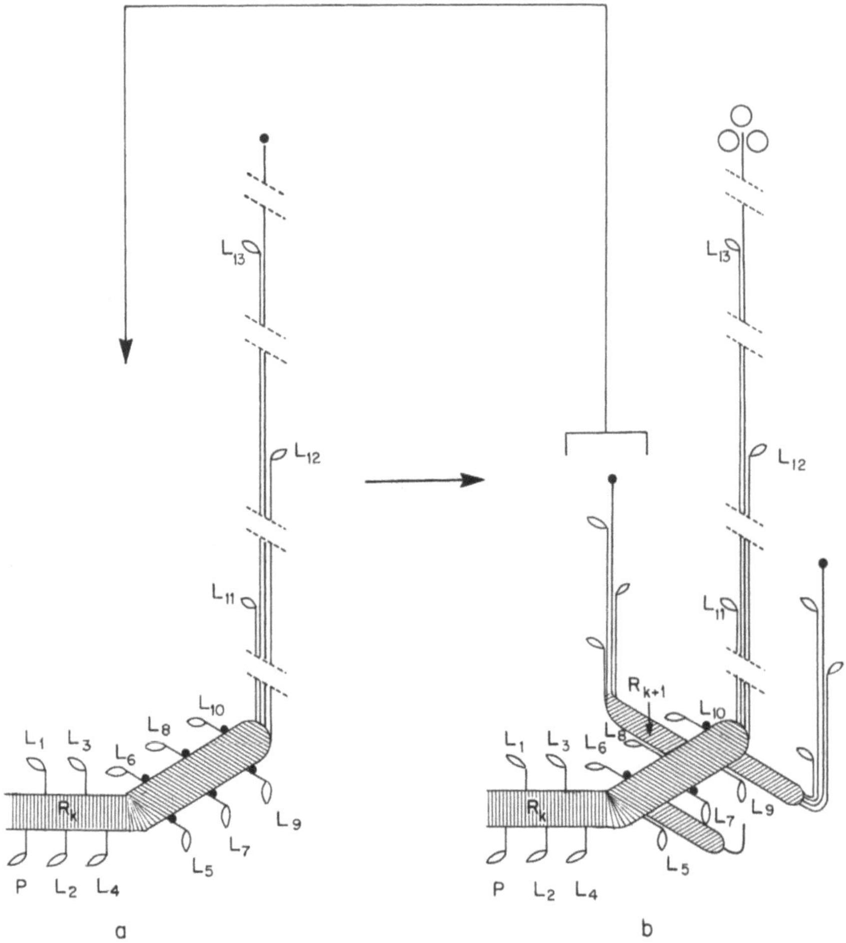

Fig. 54a,b. Schematic presentation of structure and development of *Stipagrostis ciliata*. **a** A single young ramet of the order R_k: *P prophyll*; L_1–L_{13} location of leaves in an age sequence; **b** a ramet of the order R_k with branches of the order R_{k+1} at the axil of L_5, L_8, and L_9

54a, R_k). The stem has short (1–2 mm) internodes along its 2–3 cm long leaf-carrying basal part, and terminates in an inflorescence, the culm of which has three internodes 8–30 cm long (Fig. 54a, L_{11}–L_{13}). The prophyll (Fig. 54, *P*) and the first four leaves of the brachyblastic basal part of the ramet bear no buds in their axils (Fig. 54a, *P*, L_1–L_4) whereas the upper five or six leaves (Fig. 54a, L_5–L_{10}) have axillary buds subtended by sheaths. The upper part of the prophyll (Fig. 55, *Pb*, *h*) has a hyaline portion which subtends the remainder of the stem while young. The wooly hairs along the margins of the hyaline portion (Fig. 55, *Pb*, *v*) may serve for protection from desiccation and sand abrasion. The lower part of *P* and L_1–L_3 are straw-coloured, pale-yellow, or whitish, whereas the

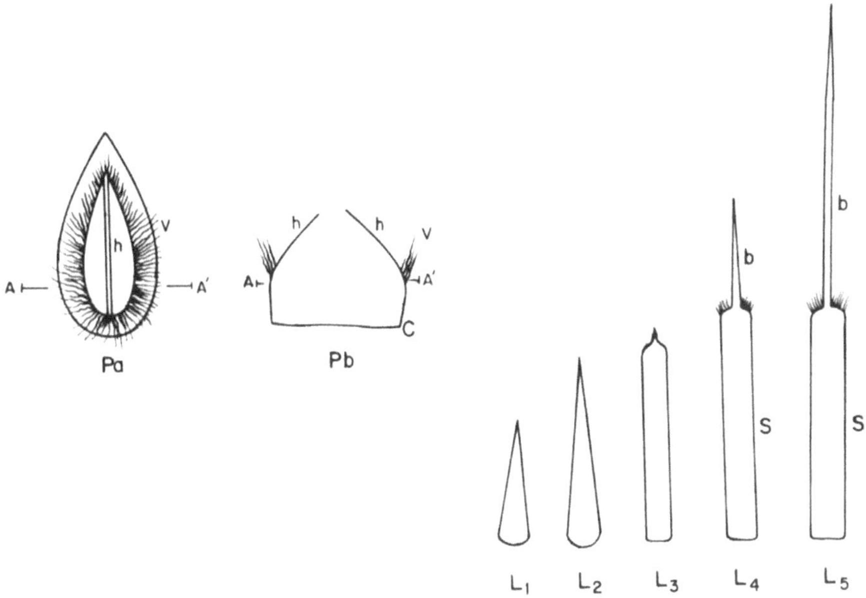

Fig. 55. Leaves of *S. ciliata*. *Pa* Rear view of the lowermost leaf (prophyll); *Pb* a transverse section of *Pa* along the line *A–A'*; *C* straw-coloured section; *h* hyaline section; *v* hairs; L_1–L_5 1st to 5th leaves; *b* blade; *S* sheath

Fig. 56. Circular tufts of *Stipagrostis ciliata* in a stable sand sheet in Negev

blades of L_4–L_{10} are green. The first leaf with a long blade is L_5. The blades of the leaves L_5–L_{13} are convoluted throughout, horseshoe-shaped in transverse section, and densely hairy on the upper surface. From amongst some of the buds at the axil of L_4–L_9, new tillers of the order R_{k+1} develop in an intravaginal pattern (Fig. 54b, R_{k+1}). Their lower brachyblastic parts are a continuation of the leafy rhizome. Due to intensive tillering a dense rounded tuft several decimeters in diameter is quickly formed. When the genet gets older its central part dies and deteriorates, leaving a dense tuft of dead stems and roots and an outer circle of satellite tufts (Fig. 56).

Similar circles of *Stipagrostis obtusa*, *Asphodelus ramosus*, *Cyperus macrorrhizus*, *Calligonum comosum*, and *Artemisia monosperma* have been observed in Israel, and of *Achnatherum hymenoides* in California. These species occur in circles in stable sand sheets. They are all capable of reproducing vegetatively. Shootborne roots are formed on each ramet which enables it to become an independent entity within the circle.

5.6.3 *Achnatherum* (= *Oryzopsis, Stipa*) hymenoides

This is a perennial grass of stable sandy soils in desert areas of southwestern USA. It has a leafy rhizome and differs in the morphology of its vegetative parts from the two *Stipagrostis* species mentioned above by the easy splitting of the tuft to single ramets. This character and the strength of the rhizome, resemble the splitting of desert shrubs into independent units (Ginzburg 1963). There are buds only on the short brachyblastic part of the ramet. The culm has 2–4 long internodes without buds in the axils of the culm leaves. However, tufts which developed on the steep slopes of nebkas created by other plants near Mono Lake and the Kelso Dunes, California, display tillering from the culm base. Sand that had slid from the upper parts of the nebkas, cover the *A. hymenoides* plants up to their second culm node. New stems with long internodes, long leaf sheaths, and nodal roots have developed from the buried nodes, enabling the plants to keep pace with the rising sand level. Adjustment of the seedling to the suitable depth is achieved by elongation of the mesocotyl (Sect. 5.8.5) and of leaf sheaths. A steppe dominated by *Artemisia tridentata* on sandy alluvium at the ancient northwestern terraces of Mono Lake contained many specimens with circles of living ramets around a center of old dead ramets, as found in the two preceding species (Sects. 5.6.1 and 5.6.2; Danin and Orshan 1995) and in *Asphodelus ramosus* (Sect. 5.6.4). The roots are covered with rhizosheaths mainly in their unbranched basipetal part, but rhizosheaths are absent at greater depth and towards the root tip. The presence of nitrogen fixing bacteria in the rhizosheaths of this plant and others was studied by Wullstein et al. (1979), while the ultramicroscopic structure of the rhizosheaths has been studied by Wullstein and Pratt (1981). The diaspore is rounded, covered by long spreading hairs, and composed of a single floret with a shiny palea and lemma. It differs considerably from those of all the other members of the genus *Achnatherum* (as presented in Hickman 1993), most of which were formerly included in the genus *Stipa*, and resembles more those of *Oryzopsis*. The diaspore may glide short distances or roll across the sand, depending on the wind velocity.

5.6.4 *Asphodelus ramosus* (= *A. microcarpus*)

This is a geophyte or hemicryptophyte of Mediterranean habitats and is well-adapted to the desert. It is found in sand sheets of the Negev in Israel. The tuft is comprised of independent ramets each with a thick leafy rhizome which is vertical in nonsandy soil and plagiotropic in sandy desert soil (Fig. 57). The following description of the plant's morphology and development is from specimens collected by me in the Negev, 10 km SE of Dimona. During summer, the leaves are dry, protecting 1–5 axillary buds. The uppermost 1–2 buds of a rhizome begin to grow in September, before the rains start. The leaves achieve their full size after the rains start. Adventitious roots then develop from meristematic tissue inside the rhizome. Later, they penetrate through the stem tissues and emerge. The young roots reach their final size during the growing season. The proximal part of the root functions as a storage organ. The distal part, up to 50 cm long, is the absorbing portion and is covered by a rhizosheath. This part may form lateral roots, and these also have rhizosheaths. The storage portion of the root may be 10–15 or 20 cm long and 1–2.5 cm in diameter. It is covered with persistet root hairs. Roots situated below the rhizome have wrikles or little folds perpendicular to the long axis. A

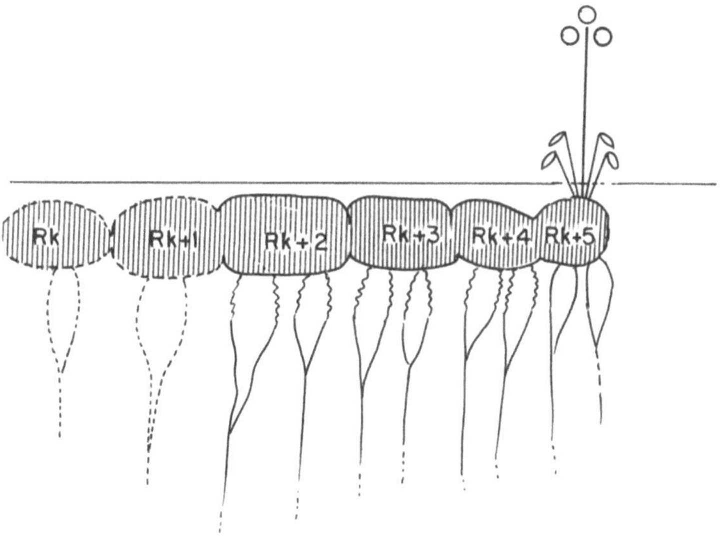

Fig. 57. Schematic presentation of structure and development of *Asphodelus ramosus* in its sixth year

longitudinal radial section of the root shows transverse white lines or strips in the parenchymatous part of the cortex and in the central cylinder. These lines occur in the proximal 0.2–0.25 portion of the root and are collapsed parenchyma cells. These roots seem to have a contractile function. Wilson and Honey (1966) concluded that root contraction is effected in many species by redirected growth of parenchyma cells, supplemented by other processes such as radial expansion and longitudinal contraction. The contraction of the roots brings the rhizome to its correct position in relation to the soil surface. The parameters for this are both depth below the soil surface and the direction of the rhizome in relation to the horizontal. The stem from which the leaves grow is perpendicular to the soil surface. As the following year's stem grows above the previous year's stem, the renewal buds would gradually become situated above the ground. Contracting roots pull the renewal buds downwards. In specimens from a sandy area near Dimona, the roots situated in a plagiotropic position have neither outer folds nor inner lines. The leafy rhizome remains alive to an age of 5–10 years in this area, and adds ca. 0.5 cm each year to the lateral growth of the plant. Rhizomes older than 5–7 years shrink with their attached roots die, and slowly disintegrate (Fig. 57 R_k and R_{k+1}). The living part of the rhizome remains unbranched, if not blooming. A year after the terminal bud has turned into an inflorescence, two buds may develop thereby leading to branching of the rhizome, and thus contributing to the circular structure of the tuft.

As a young plant grows older it branches in all directions into independent ramets and becomes a circular tuft 50–100 cm in diameter. In time, the center of the tuft dies and the plant continues growing towards the periphery, thus creating a circle of satellite tufts (Fig. 58). Circles 0.5–5 m diameter have been

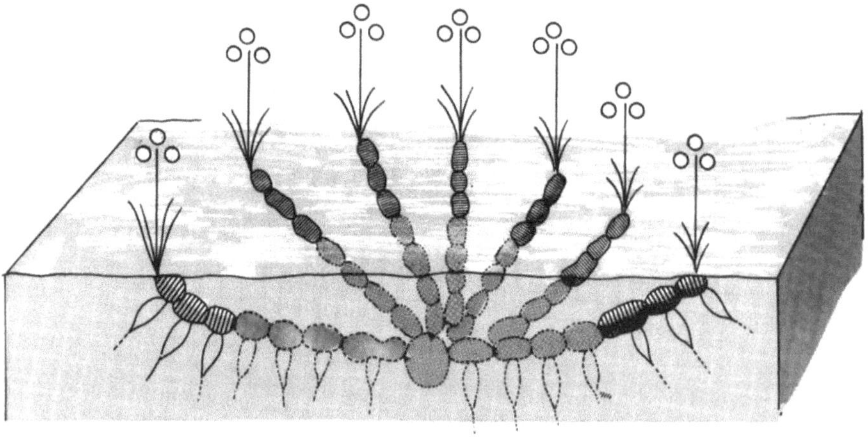

Fig. 58. A schematic representation of the structure of a circle of *A. ramosus*

Fig. 59. A circle of *A. ramosus* on a sand sheet, northern Negev

observed 10 km NE of Dimona (Fig. 59). There are large concentrations of this species in Mishor Yamin, some of them in circular clones.

Plants from the coastal dunes of Plakias, southern Crete (400–500 mm mean annual rainfall), in a site with constant sand accretion, display vertical leafy rhizomes more than 50 cm long (Fig. 60). The sequence of events leading to Fig. 60 is analyzed in Fig. 61. The response of *A. ramosus* to sand covering the renewal bud to level A (Fig. 61a,b) was negative geotropic growth of the rhizome to the 'correct' position. Additional cover to level B (Fig. 61c) was followed by additional growth of the rhizome. Sand erosion by wind to level D (Fig. 61d) then led to the final shape.

5.6.5 Summary

Plants belonging to this group share a few ecomorphological properties.

1. They seem to be sensitive to sand accretion and are rarely found in habitats where these occur. They are relatively resistant to deflation. They typically grow in stable sand sheets or stabilized dunes.
2. They tend to form dense tufts which restrict sand deposition.
3. The plants' leafy rhizome develop without disturbance branches and, in time, tend to form a circular clone of independent ramets.
4. They all have rhizosheaths.

Fig. 60. *A. ramosus* on the coastal plain of Plakias, Crete

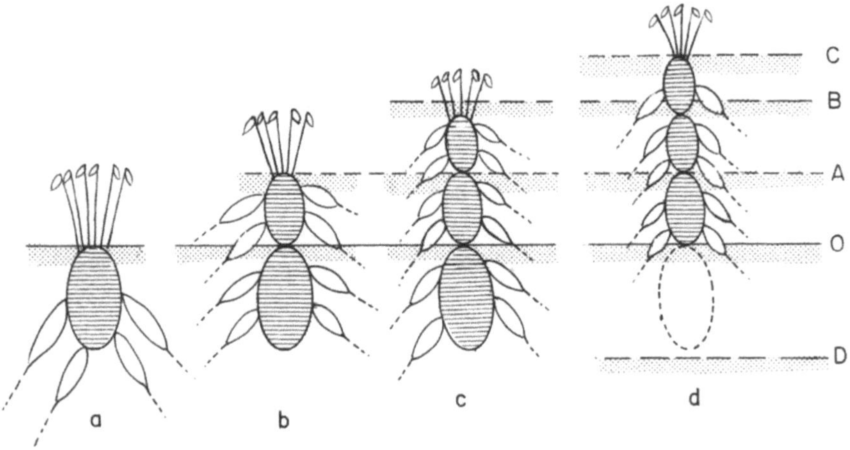

Fig. 61a–d. Schematic presentation of structure and development of a vertical leafy rhizome of *Asphodelus ramosus* from Fig. 60. *A*, *B*, and *C* are layers of sand which covered the primary layer *0*; at stage *d* all the sand was eroded to layer *D*

5.7 Shrubs and Semishrubs of Stable Sand Sheets – *Hammada salicornica* Type

Under desert conditions in Israel and Sinai, the following species all grow in stable sand sheets. In most cases they do not produce adventitious roots when covered with sand. Although many are Chenopodiaceae, this character is not restricted to that family. *Retama raetam* (Sect. 5.5.3) and other species mentioned in Section 5.5 belong here too. However, not all species growing in sand sheets have the ability to withstand extensive exposure of their roots and two groups have therefore been recognized. They differ in their demands and are consequently found as dominants in areas with different environmental conditions.

5.7.1 *Hammada salicornica* (= *Haloxylon salicornicum*)

This species is one of the most common semishrubs or shrubs of the sand deserts of the Middle East. It is common at elevations below 1000 m in Sinai and Israel. It cannot tolerate much sand accretion but it will live for a long time on stable or slightly mobile (accreting and deflating) sand sheets (Fig. 62). It frequently forms nebkas; I have never found shootborne roots in this species.

Fig. 62. Scattered shrubs of *Hammada salicornica* on a sand sheet in western Sinai, 20 km NE of Suez

Its growing season is the summer (Danin 1983), and it blooms in autumn, therefore I would expect it to have an extensive, deep root system.

5.7.2 *Anabasis articulata*

This is a plant of stable or semistable sandy soils in the Middle Eastern deserts (Danin 1983). It withstands cover by a few centimeters of sand or exposure of its roots resulting from sand deflation. When covered with sand it produces very few new roots. Exposed roots are relatively resistant to desiccation, probably due to the anatomy of the wood of the Chenopodiaceae in which bundles of xylem and phloem are scattered throughout the whole cross section of the root (Fahn and Cutler 1992). If peripheral bundles become desiccated, internal ones may continue to function.

5.7.3 *Haloxylon persicum*

In most sites in Israel, Jordan, and Sinai *H. persicum* is a shrub or low tree, 1–4 m tall. It grows in sandy soil or even on dunes in extreme desert areas with less than 50 mm mean annual rainfall. It rarely shows adaptations to sand mobility and often suffers from this. I have examined dozens of specimens in the southern Negev and the Arava of Israel, and found only three with shootborne roots. It is possible that the sand which covered most of the specimens was dry and hence did not trigger the formation of adventitious roots. This species does not form shootborne roots in the Karakum desert either (Walter and Box 1983a). It may well be that adventitious roots develop in extreme desert only on exceptional occasions when sand cover is followed by efficient rain. At Biqa'at Uvda in the southern Negev, most of the specimens covered by sand did not respond positively to burial and did not grow above the sand cover. Some specimens died as a result of being covered by sand but others survived when reexposed by deflation. The thin tips of their branches died, however, but dormant buds 15–30 cm below their tips gave rise to new branches. *H. persicum* may withstand sand cover up to 3 m depth for a limited period of time (the length of which I do not know). The roots may be exposed to a depth of 1 m and survive, provided that the entire root system is not exposed (Fig. 63). The most common habitat of the plant in Israel is sandy alluvium which receives runoff water beyond the 50 mm annual rainfall of the area. The plant's crown functions as a windbreak and considerable amounts of sand may accumulate around it. It commonly forms nebkas.

A second habitat in which it is found germinating, is mobile sand dunes. In the summer of 1993, several dead seedlings were found on the dunes of the Arava Valley. The same dunes supported well-developed seedlings in May 1994, after an exceptionally wet December. The seedling has a long taproot which seems to reach soil layers with relatively high water content. The dispersal unit is a one-seeded fruit subtended by five sepals, each armed with

Fig. 63. *Haloxylon persicum* with exposed roots in a sandy plain, Gebel Maghara, N Sinai

a lateral wing. I assume that the dispersal procedure is closer to rolling on the dune surface rather than gliding in the air. *H. ammodendron,* is the most important species of *Haloxylon* in the saksa'ul vegetation of the Karakum and Kyzilkum deserts. It is an alkali halophyte which forms dense thickets on soils with shallow groundwater tables (Walter and Box 1983a).

5.7.4 Summary

The plants belonging to this group are rather widespread semishrubs or shrubs of the sand deserts of the Middle East, Asia, and North Africa. They grow on sandy soils where the substrate is mostly stable. They may withstand cover of their stems nearly to the tips but they do not produce adventitious roots and show no obvious rapid reactive growth of their stems.

5.8 Annuals and Facultative Annuals (Amphiphytes)

Annual plants are an extremely important growth form in Mediterranean-type desert habitats (Danin and Orshan 1990) and particularly in dunes. In extreme desert areas, annuals are not seen every year. Many avoid the dry years and become active in response to a rain "pulse" (Noy-Meir 1973; Gutterman 1993).

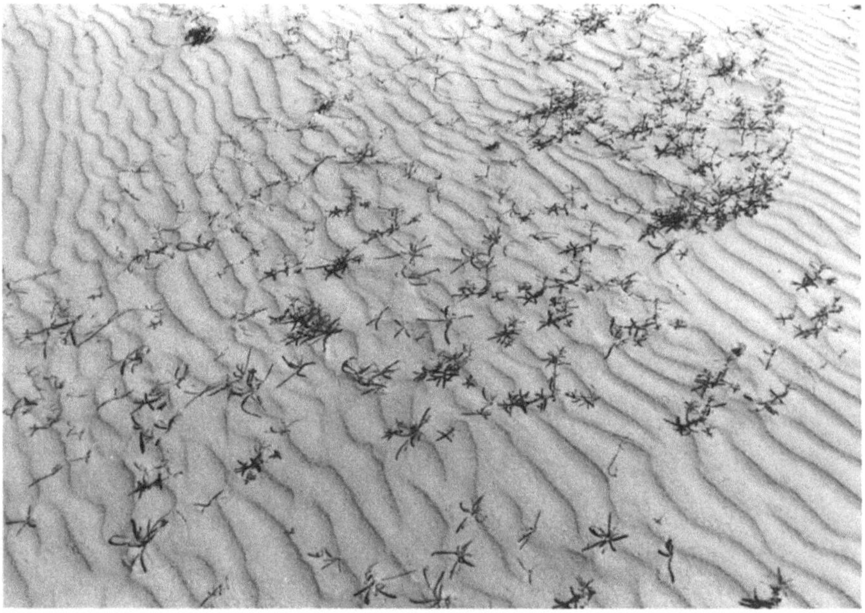

Fig. 64. Mobile sand dominated by *Eremobium aegyptiacum* in the winter of a wet year in the extreme desert area of the Arava, 60 km N of Elat

Yeaton (1988) found, in Namibia, that after "relatively heavy rains" annual plant species are to be found on the dune slopes. At least 10 mm rain, at the beginning of the season, is needed to stimulate annual germination on sand of the Arava, Israel (Naor, pers. comm.). Certain geomorphological positions, such as the leeward sides of dunes, are devoid of annuals because dry wind-blown sand grains accumulate there (Fig. 65).

Some annuals are amphiphytic (sensu Orshan 1986b). That is, they may be annual or perennial depending on local environmental conditions. Several species of *Stipagrostis* (*S. plumosa*, *S. ciliata*, and *S. hirtigluma*) are perennial under moderate conditions but are annual under extreme desert conditions. A few species of *Fagonia* and *Polycarpaea* have the same ability.

5.8.1 *Stipagrostis ciliata*

In most parts of Israel this is always a perennial plant. But, in extreme deserts as at Biqa'at Uvda and in the Arava valley, it is sometimes an annual. In Namibia, the perennial form grows in rock crevices and, in rainy years, the annual form covers the interdune and low dune slopes. This ability to become an annual, increases the plant's plasticity, thus enabling it to survive in areas where the period of water availability is too short for a perennial. The same phenomenon takes place in *S. hirtigluma* which functions as a perennial in the rocky mountains of Botswana and around Pretoria, South Africa, (specimens at Kew) and as an annual on the slopes of alluvial material near Elat, Israel. There are a few species of *Stipagrostis* which are true annuals, such as *S. namibensis* which grows on sandy soils in the extreme desert areas of Namibia only in wet years.

5.8.2 *Centropodia forskalii*

This species is most often a perennial which grows under a wide range of climatic and edaphic conditions. Annual specimens, which I collected in winter 1992–1993 on the dunes of the extreme desert of the Arava Valley, had less than ten leaves, a short inflorescence, a seminal root only, and no nodal roots. They were 5 cm tall, whereas the perennial form, growing nearby in a moister microhabitat was 70 cm high.

5.8.3 *Plantago cylindrica*

This is the most common annual psammophyte in the extreme desert sandy areas of Israel. In the Arava Valley and the southern Negev it dominates extensive areas of stable or semistable dunes. In areas of mobile dunes it is less abundant and found as a companion to *Eremobium aegyptiacum*. It remains alive and keeps functioning even when its roots are exposed to a depth of 5–10 cm or when its stems and leaves are covered by a thick layer of sand. The leaves and stems are covered with appressed hairs, contributing to its light colour. They provide protection from the abrasive effect of airborne sand grains. It has a persistent skeleton and is often buried by sand. Growing in the Arava Valley in relatively stable sand sheets from less than 5 cm to more than a meter in thickness, the plant creates its own habitat by stabilizing the sand around itself, thus enabling additional specimens to grow in the coming years. The sand sheets dominated by this species are "fortified" by dead skeletons which are not easily detached from the ground and do not deteriorate in the extremely dry desert conditions.

The seeds are 1.5–2 mm long, 1 mm wide, and 0.3–0.5 mm thick. When wetted, they exude a mucilaginous material (cf. Danin 1983; Fig 22). Although larger than sand grains, they may participate in the common movements of sand, i.e. saltation, creeping, and rolling.

5.8.4 *Eremobium aegyptiacum*

In most sites, this is an annual. It is rarely perennial and then only a biennial. The entire plant is densely covered by appressed stellate hairs. Here, again, the indumentum (type of hairs or other trichomes covering the organ's surface) probably functions both for reflection of solar radiation and for protection from injuries implicated by airborne sand grains. Well-developed specimens

Fig. 65. Mobile sand dominated by *Eremobium aegyptiacum* and stable sand by *Plantago cylindrica* (in the area of Fig. 64)

are resistant to sand cover or sand deflation up to 10–15 cm, as observed in the Arava Valley in March 1994.

The seeds are 1.5–2 mm long, 0.8–1 mm wide, 0.3–0.5 mm thick, flat on one side and rounded on the other side. They are yellow and among the sand grains are inconspicuous to granivorous animals. The penetration of seeds into a dune turns it into a seed bank. This means that the entire dune contains large quantities of seeds. Seeds distributed with mobile sand are mixed with the sand throughout its depth. When a particular layer of sand is exposed, or approaches the soil surface, the seeds in it germinate and the entire dune becomes covered with plants as observed in the Arava Valley in 1994 (Fig. 64). The leeward part of the dune (Fig. 65) is devoid of plants not because there are no seeds there but because the sand and the seeds in it are so dry.

5.8.5 *Cutandia memphitica*

This is a glabrous annual grass, one of the best-adapted annuals to mobile sand in the Negev and N Sinai deserts. Well-developed specimens are resistant to sand cover and sand deflation to a depth of 10–15 cm. When covered, the plant produces elongated internodes which assist it in escaping from the sand. In a plant community dominated by *Stipagrostis scoparia*, it is sometimes the only annual companion (Danin 1978b).

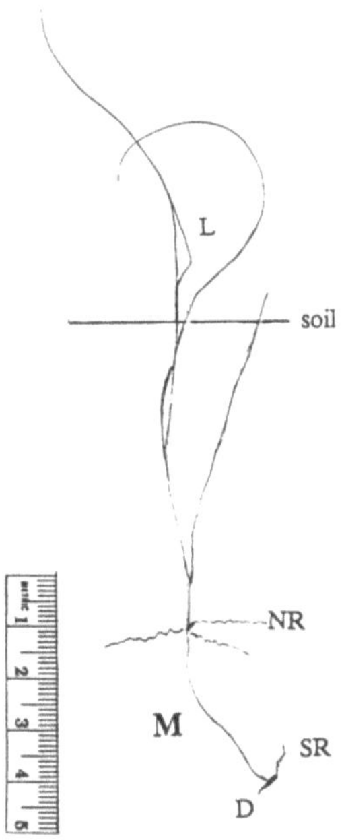

Fig. 66. *Cutandia memphitica* from a site with sand accretion, the Negev, Israel, 45 km SSW of Beer Sheva: *D* diaspore; *L* leaves; *M* mesocotyl; *NR* nodal roots; *SR* seminal roots

The diaspore includes a caryopsis and a glabrous palea and lemma. There is no apparatus which would assist dispersal. When developing in stable sand it may germinate from a depth of 1–2 cm, develop a "crown" of leaves (a section of a stem base in the grass seedling where two or more nodes remain close together, Ries and Hofmann 1991) and inflorescences, both terminal and lateral, from the axils of almost every leaf. When growing in a site of sand accretion, *Cutandia memphitica* is able to emerge above the sand surface. The mechanism of emergence and exposure of the leaf blades involves the following components: (1) elongation of the mesocotyl; (2) elongation of the first cauline internodes; and (3) elongation of the leaf sheaths (Fig. 66). The mesocotyl is the first internode between the scutellar and coleoptilar nodes (Ries and Hofmann 1991). According to Fahn (1982), the mesocotyl is a complex structure formed by the fusion of parts of the cotyledon with the hypocotyl or the first internode of the axis. The elongated mesocotyl of *C. memphitica* looks like a root without root hairs. Remnants of the scutellum (the monocot cotyledon) are easily distinguished at the transition between the mesocotyl and the seminal root. Morphological features representing the reactive growth of several organs of *C. memphitica* under sand accretion, were studied on a nebka of *Stipagrostis scoparia*, 40 km S of Beer Sheva; results are presented in Table 2. All three mechanisms of maintaining the leaf blades above the sand surface are exhibited. However, the nonlinear increase with sand depth in each of the five parameters measured implies that they may function independently at various ages of the plant. This issue deserves further

Table 2. Average vegetative growth parameters of *Cutandia memphitica* plants (mm) derived from a slope of a nebka, 40 km south of Beer Sheva, Israel

Sand depth (mm)	Mesocotyl length (mm)	Length of internodes 1–3 (mm)	Length of leaf sheath (mm)		
			L_1	L_2	L_3
10	1.4 ± 1.0	0.4 ± 0.5	7.0 ± 2.1	8.0 ± 2.1	9.0 ± 1.1
20–30	19.5 ± 12.6	18.7 ± 4.6	9.7 ± 1.1	19.3 ± 3.0	19.0 ± 4.2
50–60	32.0 ± 7.3	35.7 ± 5.9	29.2 ± 2.5	39.8 ± 4.3	35.3 ± 12.8
60–70	30.2 ± 6.0	28.3 ± 5.7	32.3 ± 3.9	43.7 ± 2.5	50.7 ± 2.4
90	14.2 ± 5.3	34.4 ± 6.6	68.4 ± 5.2	67.4 ± 7.1	64.8 ± 4.8
100	35.3 ± 23.0	42.0 ± 7.1	38.3 ± 4.7	43.7 ± 7.0	50.7 ± 4.2
120	30.5 ± 6.3	86.3 ± 28.1	34.3 ± 7.0	46.3 ± 14.5	58.0 ± 11.0
140	26.3 ± 3.4	42.3 ± 29.2	63.0 ± 0.0	71.0 ± 8.3	63.3 ± 21.6

Sand depth = above the first nodal root; L_1, L_2, L_3 = first through third leaves; ± = standard deviation

experimental work, in particular to resolve questions concerning light and temperature sensitivity. Ries and Hofmann (1991) found, in one species, that the reactive growth resulting from deep sowing was elongation of the mesocotyl. The mesocotyl of the other two species did not elongate. At least one dicotyledonous species, *Dicoria canescens*, emerges in a similar way by elongation of the hypocotyl (Sect. 5.10.6). The elongation of the three organs is probably a result of the activity of intercalary meristems. Elongation of leaf sheaths and of stem internodes are two of the modes by which adult plants of *Panicum urvilleanum* (Sect. 5.1.7) withstands sand accretion.

5.8.6 Summary

The annuals are a heterogenic group. The few traits they share are:

1. Renewal buds are protected in seeds which have a high diversity of protection measures.
2. Germination is facilitated only under conditions which will permit completion of the life cycle. Burial of diaspores in the sand may be a limiting factor for germination.
3. Several annuals display resistance to sand accretion and deflation. Some of them may show reactive elongation of internodes after burial by sand. One strictly annual grass has the ability to adjust to sand accretion by elongation of the mesocotyl and leaf sheaths.
4. Several species are facultative annuals and become perennial in wetter years or on nonsandy substrates.
5. In several species, seed dispersal takes place as part of sand movement. Some species have sand coloured seeds which may decrease the risk of predation.

5.9 Coastal Dune Plants

Many coastal plants are adapted to withstand the hazards of the coast. The most serious of these is the constant spray of sea water. The plants studied here were as far from the actual beach line as possible. They represent adaptations to coastal sand, although some of their physiological adaptations may be associated with a different nutrient regime near the coast. However, most of the ecomorphological properties discussed here are concerned with sand mobility rather than with the nutrient situation.

5.9.1 *Ammophila arenaria*

This is a robust and sclerophyllous perennial grass which functions as an evergreen geophyte or chamaephyte. It is confined to mobile sand dunes, mostly in coastal areas of Europe, Asia, and Africa (Huiskes 1979; Gehu 1985). It was introduced to the western coast of the USA in the late 1800s. It has since spread north to Canada and south to San Louis Obispo (Breckon and Barbour 1974). Environmentalists agree that the plant spreads aggressively in California and in many places has caused the disappearance of the native vegetation of the mobile dunes. It occupies mobile sand, but may also dominate stable coastal dunes in Europe ("grey dunes" of Chapman 1976). Old stable coastal dunes at Bodega Head and Point Reyes peninsula, California, for example, are dominated by *A. arenaria.*

A tuft which looks caespitose or a tussock (Greig-Smith et al. 1947) may be composed of several ramets, each having the following morphology. An extravaginal tiller of the parent plant of the order R_k is a scaly rhizode with long internodes. Its distal part is brachyblastic, 1 cm long, carrying 5–10 leaves; it may terminate with an inflorescence, thus making the ramet a sympodium. There is a bud in the axil of each rhizode scale and in the axils of the lowest leaves of the brachyblastic parts of the tillers. There are 3–4 root primordia at each node. As the buds of the terminal part of the ramet start to grow, they give rise to 3–5 tillers of the order R_{k+1}. Each of these branches into 3–5 tillers of the order R_{k+2} which, in turn, have 2–3 tillers (R_{k+3}). Each of the R_{k+3} tillers has 5–10 leaves, making the leaf density of the whole tuft rather high. As a result, each main branch of the plant looks like a broom at the end of a rhizode. When a branched stem is covered by sand, it develops roots from the nodal primordia and its buds give rise to tillers at a higher level. Thus, the aerial tillers become orthogeocorms. On stable dunes in Bodega, California, the internodes of aerial tillers were 0.2–0.5 cm long in spring. They were 2–25 cm long on the windward slope of the foredune where sand accretion is most intensive. The leaf sheaths in both habitats were about the same length i.e., 15–25 cm long. If not covered, new roots do not form, the stems and leaves senescence, and the plant dies. Studies on growth and development demonstrating the adaptations of grasses to sand accretion on the coastal dunes of Europe are reported by Marshall

(1965) on *Corynephorus canescens*, and by Crawford (1989) on *Ammophila arenaria*. In each case the root system rapidly senesces and the plant requires a fresh deposition of sand around the shoot to encourage the growth of new nodal roots. Occasional buds on an old, senescent tiller may sprout and extend the life of the tuft, either as an intravaginal orthogeocorm or as an extravaginal rhizode. Such reactive growth can maintain the population unless other plants enter the area and outcompete the *A. arenaria*.

Ammophila arenaria reproduces by the extravaginal tillering of rhizodes in a nonrhythmic way; the rhizodes develop occasionally, not as part of the regular growth of each ramet but mainly by the vegetative reproduction of old tufts. The rhizode internodes may be 10–25 cm long, whereas the internodes at the branching or tillering tip of the rhizode are only 0.5–2 cm long. Most of the growth of *Ammophila arenaria* in Israel and in California is by intravaginal tillering of the brachyblastic part of the main stem of the seedling and of its extravaginal rhizodic tillers. As a result, an area dominated by *Ammophila arenaria* exhibits a group of tufts rather than a web of small plants in rows (Sect. 5.9.2). In a detailed aerial photograph, a young dune dominated by *A. arenaria* may look like a star patch. The tuft, which develops from a seedling gives rise to rhizodes which produce vertical tillers. In time, each tiller expands, resulting in easily recognized "star" arms. As previously stated by Gemmell et al. (1953), the tillers appear to develop more frequently on the vertical parts of the ramet than along the horizontal rhizode. Growing in areas with relatively high amounts of rainfall, the continuous vertical reactive growth of orthogeocorms leads to the formation of high dunes. This is true in many coastal areas of Europe where the foredune looks like a vegetated wall separating the land from the sea. The tendency for the foredune to grow vertically and to arrest mobile sand in coastal California has been discussed by Barbour et al. (1993) in relation to its impact on decreasing habitat and species diversity. Our study of plant succession on coastal dunes in Bodega Head, California (Danin et al. 1995) shows sand accretion in *A. arenaria* nebkas to have been at a rate of 10 cm/year for the last 40 years.

The diaspore is a floret with a palea and lemma which have bristly hairs at the tip and shorter ones throughout. All the hairs are turned upward. The diaspores are poorly adapted for wind dispersal. Their modes of dispersal resemble those of the desert species, described in Section 5.1, which require sand mobility for survival. Further observations and experimental work will be needed to determine if dispersal is mainly by "gliding" or "rolling".

5.9.2 *Ammophila breviligulata*

This is a perennial grass which functions as an evergreen geophyte. It is confined to mobile or semistable sand dunes, mostly in coastal areas of northeastern USA and eastern Canada. *A. breviligulata* has a well-developed system of scaly rhizodes covered with many buds, a few of which sprout intravaginally and give rise to short vertical tillers. Each tiller has a brachyblastic portion with a few axillary buds which tend to become smaller on higher positions. One bud and 3–4 primordia of roots occur at each node. The inflorescence culm has 1–2 nodes. I have observed *A. breviligulata* at Napeague Bay dunes in eastern Long Island: individuals on a mobile dune looked caespitose and similar to *A. arenaria*. However, individuals found on a rather more stable dune were densely propagated from rhizodes. All the tillers of the order R_{k+1}, sprouting from rhizodes (order R_k), had small diameters and were arranged in rows. This means that vertical growth is limited as compared to the lateral growth. Unlike that of *A. arenaria*, the vertical tillers of the rhizode only attain the order of R_{k+1} and do not ramify further.

The main "investment" of A. *arenaria* is in vertical and not in lateral growth. Consequently, a dune dominated by *A. breviligulata* looks like a loose green carpet with thin ramets, each with 3–10 fully developed leaves, without nebkas around the individual ramets. The entire dune surface appears smooth. In contrast, a dune dominated by *A. arenaria* has an uneven topography because each individual forms a nebka - sometimes a very large nebka (Hesp 1981; Danin and Nokrian 1991). An area densely covered by *A. arenaria* is always 1–3 (–5) m higher than the nonvegetated surface of the dune. Each individual plant (one genet) has dozens to hundreds of leaves arranged on tillers which are connected by a leafy rhizome. Only a few leaves in such a genet are carried on rhizodes. A genet of *A. breviligulata*, on the other hand, has most of its leaves on long rhizodes. The bud bank of *A. breviligulata* is mainly on rhizodes below the ground, whereas the majority of buds in *A. arenaria* is on vertical tillers, many of which are above the ground.

Ammophila breviligulata has been studied in the field by Disraeli (1984) who showed that vigor increased markedly when the plant was buried by a layer of sand 2–35 cm deep. Vigor was demonstrated by increases in both below ground and above ground biomass. Maximal leaf area occurred when buried in sand to a depth of 22 cm. Plant height, the number of buds per tiller, the number of rhizomes per plant, internode length, and chlorophyll concentration, all increased with burial in sand up to a certain threshold. Plants in sites without sand accretion showed reduced vigor. Disraeli assumed vigor to be related to the formation of new roots from the buried stems. Old roots are not able to absorb nutrients as efficiently as younger ones. Experimental work carried out by Seliskar (1994) showed that sand accretion affected the carbohydrate reserves in the rhizodes.

5.9.3 *Spartina patens*

This is a geophytic or hemicryptophytic perennial grass. It is confined mainly to wet salt marshes in North America, but occasionally grows on the coastal foredunes of beach slopes facing the sea. In eastern Long Island it forms monospecific patches on the sands of the foredunes. *S. patens* is preadapted to mobile sands by its rhizodes, which are rich in axillary buds. There are buds on the vertical stems as well, but the subterranean buds are larger and more vigorous. Bud vitality appears to decrease in an acropetal sequence. Although successful on sea-facing sand, it is rare on dunes only a few hundred meters inland. It can be regarded as a facultative plant of the shore or the foredune.

5.9.4 *Elymus farctus*

This is a perennial grass of Mediterranean and European beaches. In its growth habits it resembles the *Stipagrostis plumosa* type of the desert dune plants (see Sect. 5.3.1); it is resistant to sand cover, sand deflation, and stable sand. In Israel it is restricted to the shore where it may grow on calcareous sandstone and form nebkas up to 1 m high. In the Antalya - Alanya area, of southern Turkey, *E. farctus* is a foredune plant. Near the city of Side, for example, it is the dominant plant on dunes where there is some sand accretion, a few hundred meters away from the shore. *E. farctus* has a rather dense silhouette (Fig. 4) which protects the plant from a high rate of sand accumulation within the tuft. The roots have rhizosheaths.

5.9.5 *Cyperus capitatus* (= *C. mucronatus*)

This is a plant of stable sand sheets and sandy loam soils in the Mediterranean region. It is 10–20 cm in height and has convoluted sclerophyllous leaves with dorsoventral anatomy. The chlorenchyma and conductive tissues are in the abaxial side of the leaf, whereas there are only parenchyma cells in the adaxial. The abaxial epidermis is thicker than the adaxial. In Israel it has thin (0.9–1.3 mm diameter) and sometimes very long scaly rhizodes with few axillary buds or roots except at its terminal part where a tuberous new vertical stem develops. This long monopodial tiller fits Du Rietz's (1931) definition of "rhizode" and the brachyblastic tuberous termination of the rhizode fits his definition of "subole". This entire organ has a sympodial branching mode. The tiller grows first as a monopodium and at a certain point becomes a sympodium. The same species in the coastal sands of southern Turkey has thicker rhizodes, at a depth of 10–30 cm, with buds and roots along the entire dolychoblastic portion. In Israeli populations, rhizodes develop closer to the soil surface. The plants' main habitat is immobile sand but, when covered by a thin layer of sand, it is capable of limited vertical growth by elongation of the internodes of the vertical stems below or above the tuber. It seems to be more mesophytic than is *Cyperus macrorrhizus* (Sect. 5.3.5), as indicated by its geographical distribution and habitats.

The ramet of this species (Fig. 67a), resembles that of *Cyperus macrorrhizus* (Fig. 35a). At the distal part of a leafy rhizome, *C. capitatus* has a vertical tuberous brachyblastic portion of 5–6 (–12) internodes. The inflorescence is borne on a culm consisting of one internode. Roots develop from the

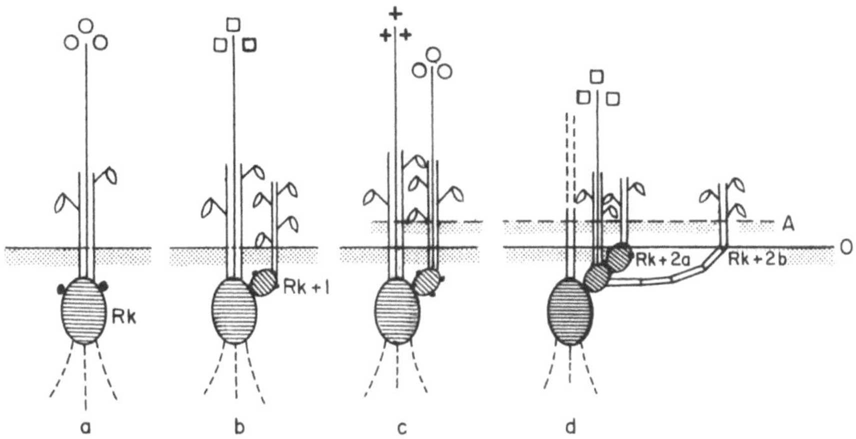

Fig. 67a–d. Schematic presentation of structure and development of *Cyperus capitatus*. **a** A flowering ramet of the R_k order with a bud at the axil of a few leaves of the tuberous brachyblastic part; **b** R_k ramet with a branch and a new ramet (R_{k+1}) developed from a bud of R_k; **c** as b, older with *A* layer of sand; **d** sprouting of R_{k+2a} (as in Fig. 30c) and of the rhizode R_{k+2b}

lower nodes of the previous year's tuber. The buds at the tuber's leaf axil may develop into tillers with tuberous basal parts that continue as a leafy rhizome (Fig. 67b). In this respect, the Israeli populations of this species differ from Turkish ones in which nearly all the vertical stems come out of the end of a rhizode. Other buds on the tuber may, by extravaginal growth, give rise to scaly rhizodes more than 50 cm long. They terminate in one or more tuberous vertical tillers, thus starting a new tuft (Fig. 67d). When covered with sand, the tuberous tiller which continues as a leafy rhizome may sprout, developing long internodes below the tuber (as described for *C. macrorrhizus*).

5.9.6 *Cyperus sharonensis*

This geophytic evergreen of the coastal sand dunes of Israel is a new species discovered during my field study in preparation for this book. Nutlet epidermal morphology resembles that of *Cyperus capitatus*. However, the species differs in its morphology, habitat, and reactive morphology. *C. sharonensis* is a robust plant (Fig. 68) with two types of leaves: narrow ones in the lower part of the stem, and wide, longer ones nearest to the developing inflorescence. It has much thicker rhizodes, with many more axillary buds, than has *C. capitatus*. The new ramet develops in most cases not as an intravaginal tiller or a continuation of the leafy rhizome (as described in *C. capitatus*), but as a terminating sympodial growth of the tuberous end of an extravaginal rhizode. When covered by sand, the rhizodes change the angle of their growth slightly and adjust to the new soil surface. Having long leaves, the plant can withstand a very thick layer of sand cover (Fig. 68). Its loose silhouette does not lead to sand accretion.

Fig. 68. *Cyperus sharonensis*, a coastal dune plant near Caesarea, Israel

5.9.7 *Stipagrostis* (= *Aristida*) *lanata*

This plant is similar to *S. plumosa* and the two may be vicariants. Apart from the fact that, in *S. lanata*, the two short lateral awns are hairy (they are glabrous in *S. plumosa* Fig. 23b), it is difficult to distinguish between the two. Sprouting may occur from buds at the base of the vertical stems, forming a continuation of the leafy rhizome (as in Fig. 31). Alternatively, sprouting can occur from aerial stems that become covered with sand (Fig. 31d). The plant may be resistant to being covered by sand to a depth of 10–15 cm. The roots have rhizosheaths.

5.9.8 *Abronia maritima*

This is a typical perennial succulent of foredunes in California (Cooper 1936; Johnson 1977). All the aboveground parts are densely covered with colleters (glandular trichomes which secrete a sticky substance). Sand adheres to these surfaces, even after burial. This plant is one of the first colonizers of the coastal shore, and is capable of building nebkas to 2 m high. Adventitious roots develop from shoots that are buried in sand.

The diaspore is a heavy, winged fruit which may roll on the ground. It floats well on water, a mode of dispersal common to other coastal plants such as *Cakile maritima* and *Pancratium maritimum*. By drifting in currents of seawater, these species can and have crossed long distances. When thrown by the waves onto the beach, they roll inland and remain in the sand dune system. *Abronia umbellata* is a northern vicariant and *A. villosa* grows on desert dunes. Both also are viscid, and sand grains adhere to their leaves and stems.

5.9.9 Herbaceous Perennial Species

Centaurea aegialophila is a hemicryptophyte of coastal dunes in southern Turkey. In summer, hairy aboveground portions become dry, and then die. It has a thick root and a short stem which produces a basal leaf rosette. If sand covers the plant, the short stem elongates and brings the renewal buds to their proper place in relation to the soil surface. Additional sand cover will, after a year or more, induce another such elongation. After deflation, exposed roots produce thick, corky bark, in contrast to that of buried roots which are smooth.

The hemicryptophyte *Chondrilla juncea*, which grows in the vicinity of *Centaurea aegialophila* but in more stable sand, has a fixed morphology. The specimens I have examined did not display any reactive growth.

Calystegia (= *Convolvulus*) *soldanella* is a prostrate, herbaceous plant with fleshy (rarely lignified) horizontal stems which arise from a thick and fleshy root 1–4 cm in diameter. Leaf blades, petioles, and stems are covered with short colleters to which sand grains adhere. When covered by sand the horizontal stems produce shootborne roots covered by rhizosheaths (cf. Sect. 7.3.4.4.2).

5.9.10 Shrubs and Trees

Shrubs and trees of nondesert coastal areas are vulnerable to being covered by shifting dunes. Many of them grow mainly in nonsandy habitats. Their coastal forms however, have the ability to produce adventitious roots from the thick lignified stems when these become buried by sand. This phenomenon is very pronounced in southern Turkey. It is exhibited by *Sarcopoterium* (= *Poterium*) *spinosum*, *Coridothymus* (= *Thymus*) *capitatus*, *Satureja thymbra*, *Pistacia lentiscus*, *Pistacia terebinthus*, and *Rhamnus lycioides*, and in Israel by *Sarcopoterium spinosum*, *Coridothymus capitatus*, and *Pistacia lentiscus*. *Prunus maritima* and *P. serotina* occur in eastern Long Island, New York, in woodlands where they are liable to be covered with sand. They also form shootborne roots. *Baccharis pilularis*, which is very common on nonsandy ground of mesic parts of California, grows on mobile coastal dunes. In such places, it may produce nebkas. It too, has the ability to produce shootborne roots.

5.9.11 Summary

The coastal dune plants listed above parallel most of the ecomorphological types and reactive growth traits found in desert habitats.

1. Coastal dune grasses do not extend inland. They only occupy mobile or recently stabilized dunes. In California *Ammophila arenaria* keeps growing even after dune stabilization.
2. The three grasses examined have rhizodes. They all have rolled leaves with a thick cuticular layer on the abaxial, outfacing side of the lamina.
3. They all have the ability to grow vertically above encroaching sand and to produce shootborne roots on orthogeocorms. They have large bud banks and may produce tillers under a wide range of environmental conditions.
4. Many coastal grasses do not have rhizosheaths. *Stipagrostis lanata* however, does have rhizosheaths as well as many of the other adaptive properties found in its desert relatives.
5. The two coastal Mediterranean *Cyperus* species differ from desert species mainly by possessing rhizodes and lacking rhizosheaths.
6. Those trees and shrubs which are typical maquis and phrygana (batha) species of the Mediterranean area, but also grow in coastal dunes, are capable of producing adventitious roots from thick and lignified branches.

5.10 Sand Grains Coating Leaves and Stems

Several of the species discussed in Sections 5.1 to 5.9 possess glandular trichomes which secrete a sticky substance that sticks sand grains to the leaf, stem, and other portions of the plant. A few examples of common desert and nondesert coastal plants which share this property are listed below. It is regarded as a protective measure, an adaptation against abrasion by airborne sand particles.

5.10.1 *Savignia parviflora*

This species is an annual of extreme desert areas in the Negev, the Arava Valley, and in Sinai. *S. parviflora* germinates in wet years and in locally mesic habitats, such as wadis. It becomes dominant where sand, 10–15 cm thick, covers limestone slopes in the northeastern part of Biqa'at Uvda, in the southern Negev. A seedling totally covered by sand (Fig. 69) shows grains on the

Fig. 69. A seedling of *Savignia parviflora* from the southern Negev; the leaves, petioles, and hypocotyl are coated with sand grains

hypocotyl, cotyledons, leaf lamina, and petiole. The sand layer protects the seedling from the abrasive action of windborne sand.

5.10.2 *Ifloga spicata*

This small annual plant, 5–10 cm high, may become a dominant in the Negev desert dunes during one of the stages of plant succession (Danin 1978b). It has a viscid leaf surface and, when growing on desert or coastal sand always has sand grains adhering to the leaves. It has two kinds of achenes. The receptacular achenes have a pappus of a few hairs which I assume does not function as an efficient gliding apparatus. These achenes are subtended by the involucral bracts. The other achenes are solitary. They are found in the axil of each involucral bract and have no pappus. The axillary solitary achenes at least may well be dispersed, together with the moving sand grains, as discussed in the case of *Eremobium aegyptiacum* and *Plantago cylindrica*. Experimental study is needed to verify these assumptions.

5.10.3 *Ononis serrata* and Other *Ononis* Species

Ononis serrata is common annual companion of several plant communities confined to sand sheets or slightly mobile sand in desert and coastal areas of the Mediterranean region. It has colleters at the leaf and stem surface; most specimens collected in sandy areas have sand grains affixed to a varying extent. Other species of *Ononis* have this kind of indumentum too as well as sand affixed to the leaves and stems. Such are *O. natrix* subsp. *stenophylla* and *O. variegata* when growing on sandy soils.

5.10.4 *Fagonia glutinosa* and Other *Fagonia* Species

All species of *Fagonia* growing on sandy ground in the Negev and Sinai are covered with sand. The nature of the colleters in *Fagonia mollis* is discussed by Fahn and Cutler (1992) who relate their role to decreasing hazards of solar radiation (see also Sect. 7.3.3.2). It is difficult to find any grain cover on plants of the latter species when they are not growing in a sandy area. The missing sand cover on the leaves of this plant in desert, but not in sandy regions may support the hypothesis that the sand cover helps resisting abrasion. *Fagonia glutinosa* is the most common species of this genus in the sandy desert areas of Israel and Sinai which is named for this important property. Being an amphyphitic plant of extreme desert areas it may function as an annual or as a perennial. Seedlings found in the southern Negev display a thick mantle of sand grains (Fig. 70) on the cotyledons, leaves, petioles, and young stems. Perennial individuals display the same habit as well. In the same area and in the Arava Valley other species such as *Fagonia bruguieri*, *F. sinaica*, and *F. arabica* likewise have sand grains attached to the leaf surfaces.

5.10.5 *Silene villosa*

This is a common annual of plant communities in sand sheets of desert and coastal areas in Israel and Sinai. All the leaves, stems, and calyces are covered with sand grains. Other species of *Silene*, are also covered with colleters, but only a few are confined to regions of sand. One such is *Silene succulenta*, a common perennial of the coastal dunes, especially those facing the sea. A relatively rare annual is *S. sedoides* which in the coastal area of northern Israel is confined to sandstone rocks of the strand community. When growing in sites where sand particles are available, this species is covered with sand too. *Silene arabica* and *Gypsophylla viscosa*, both desert psammophytes, are also covered by sand.

5.10.6 *Dicoria canescens*

This is a whitish hairy summer annual of dunes in southwestern USA (Pavlik 1985; Hickman 1993). It also grows on the sides of roads where there is a sandy

Fig. 70. A seedling of *Fagonia glutinosa* from the southern Negev; all young plant parts are coated with sand grains

component in the soil. Seedlings collected in Eureka Dunes and the Kelso Dunes in March and April display an interesting adaptation. The hypocotyl is covered by viscid material and adhered sand grains. Appressed stiff hairs cover the epicotyl. The length of the hypocotyl is 3–10 cm and that of the epicotyl 0.3–3 cm. The ability of these two to elongate is an efficient adaptation to desert dunes where the depth of the seed bank may vary. The process resembles depth adjustment by the mesocotyl in *Cutandia memphitica* (Sect. 5.8.5). Scattered trichomes, which look like root hairs, occur on the buried hypocotyls of seedlings of *D. canescens.* After germination, the seedling may have to withstand the abrasive effect of airborne sand. Exposed seedlings of *Dicoria canescens* have the advantage that the subterranean portion of the hypocotyl is protected from abrasion by affixed sand. The base of the two cotyledons is united in such a way that the apex of the young stem and the cotyledonary buds are protected. These buds are well developed in sites of sand accretion.

5.10.7 Summary

The property of possessing viscid material or colleters affixing sand grains is found in desert and coastal psammophytes. It is not restricted to plants of dry habitats and may be seen on the aboveground and superficial parts of the plants.

5.11 Microbiotic Crust

Specialists who study the microorganisms composing the microbiotic crust of the desert may not be ready to look at the role of these organisms in the ecosystem unless they can determine their identity to species or genus level. There are, however many reasons for studying the role of the components of the microbiotic community, even if these can only be determined at the general group level. Such an approach is often practiced in the study of communities of higher plants, when physiognomic terms such as forest, maquis, scrub steppe, or heath are used. The groups of microorganisms that merit study include filamentous cyanobacteria, coccoid cyanobacteria, cushion mosses, cyanophilous lichens, etc.

5.11.1 Cyanobacteria and Green Algae

Filamentous cyanobacteria may develop on the sand surface (Danin 1978b, 1987; de Winder et al. 1989) or slightly below it (Friedmann and Galun 1974; Fig. 2). When growing in the upper layer these cyanobacteria form a crust composed of aggregated soil particles held together by entangled filaments of the cyanobacteria or glued together by the gelatinous sheath excreted by their trichomes. The crust may also include free-living fungi (Fletcher and Martin 1948; Friedmann and Galun 1974). Such a crust may induce a series of processes that lead to further stabilization of the sand, increasing its water-holding capacity and protecting the soil from further erosion (Booth 1941; Fletcher and Martin 1948; Danin 1987; Danin et al. 1989). Reviews of the cyanobacteria living in the soils of desert and semi-desert areas are presented by Friedmann and Galun (1974), Starks et al. (1981) Harper and Marble (1988), West (1990), and Johansen (1993). In many cases, the crusts resemble a network of filaments (Fig. 12) and threads of mucilage (Fig. 11) attached to the soil particles. As a result, when the crust is broken, cyanobacterial filaments coated by adhering grains of sand become visible (Fig. 10). The results of digital mapping of Ca, Si, Al, and Fe lead to the conclusion that cyanobacterial filaments are covered mainly by silt-sized and clay-sized particles of carbonates, clays, and quartz (Danin et al. 1989).

The most common filamentous cyanobacterial component in the crusts of all the sites studied by Danin et al. (1989) is *Microcoleus* sp. *M. vaginatus* is believed to be the most widespread species of cyanobacteria in the microbiotic crusts of the arid and semiarid regions of the world (Johansen 1993). It is accompanied in the Negev by species of *Nostoc*, *Schizothrix*, *Scytonema*, *Calothrix*, *Chroococcidiopsis*, and *Phormidium*. In a later study (Danin and Dor, unpublished data), thalli of the cyanophilous lichen *Collema* sp. were also seen to form part of the crust.

The dominance of cyanobacteria as pioneer colonizers in these areas derives from their ability to interact with soil particles and form wind-resistant mats. Resistance to low water potential (Brock 1975), and high temperature (Buzer et al. 1985), and the avoidance of solar radiation by growing below or among soil particles that decrease intensity (Levy and Steinberger 1986), ensure the survival of these mats during the long dry periods between the rains. Cyanobacteria in young crusts are green. Many of the species involved are filamentous, and found below a thin protective layer of sand or silt. In time, succession proceeds and other species, not necessarily filamentous, develop above ground (Danin et al. 1989) They are coloured dark by pigments that protect the cells from photooxidative death (Abeliovich and Shilo 1972). Dark soils with a microbiotic crust absorb solar radiation more efficiently and hence become warmer than nearby soils without crusts (Harper and Marble 1988; Harper and Pendleton 1993).

Bar-Or and Danin (1989) propose several mechanisms that contribute to the process of soil particle aggregation by cyanobacteria. These are: (1) mechanical binding of particles entangled by a three dimensional network of filaments; (2) adhesion to the mucilaginous sheath or slime layer produced by several species of cyanobacteria in the crust; (3) attachment of particles to sites along the cyanobacterial cell-walls. Adhesion and co-flocculation of sedimentary clay particles with cyanobacterial filaments has also been found in aquatic species (Bar-Or and Shilo 1988). Adsorption of the clay is facilitated by cell-bound extracellular polysaccharidic flocculants (Bar-Or and Shilo 1987). It also leads to the formation of cyanobacterial mats in the benthos. Similar mechanisms may be involved in the aggregation of soil particles by terrestrial cyanobacteria.

The increasing stability of the sand promotes the development of higher plants which may influence further sedimentation and the trapping of silt and clay by locally decreasing wind velocity (Danin 1978b). The amelioration of the moisture regime during plant succession is accompanied by the establishment of additional species of cyanobacteria. *Nostoc* spp. are known as nitrogen fixers, and their establishment may contribute to the improvement of the nutrient regime (Fletcher and Martin 1948; Fuller et al. 1960; Mayland et al. 1966; Stewart 1967; Rychert and Skujins 1974; Loeftis and Kurtz 1980; Johansen 1993). The amounts of available phosphorus, exchangeable potassium, and organic matter at the soil surface increase during development of the components of the microbiotic crust (Kleiner and Harper 1977; Anderson et al. 1982; Danin et al. 1989). In addition to N, P, K, Harper and Pendleton (1993) found greater amounts of Ca, Mg, and Mn in crusted, as compared with uncrusted soils. The sheath materials of cyanobacteria include chelating agents which increase the availability of various nutrients in the soil, not only for the cyanobacteria but for many other components of the ecosystem (Lange 1974). The uptake of elements, such as N, P, K, Ca, Mg, Mn, Zn, Fe, Cu, and Na, by seed plants is influenced by edaphic cyanobacteria (Harper and Pendleton 1993).

Populating the entire soil surface, the cyanobacteria with their polysaccharide sheaths induce the formation of cracks through their impact on the level of shrinkage and imbibition. This ability is derived from the physical properties of their mucilaginous sheaths (Navarini et al. 1992). Hence, crusts of increasing age on stable sand dunes turn into cracked surfaces because of the trapping of fine-grained particles and the increasing quantities of polysaccharide sheath material in the crust.

When comparing thallus structures of the species prevailing in the early and late stages of succession, it becomes evident that entirely different life strategies are displayed. *Microcoleus* and *Schizothrix*, which are flat network-forming filamentous cyanobacteria, dominate the first stages of microbiotic colonization. *Microcoleus vaginatus* shows pronounced motility, which helps it to emerge above the covering sediment (Dor 1995). These two organisms are endedaphic (i.e., they grow inside the soil; Friedmann and Galun 1974) whereas the cyanobacterium *Nostoc* sp. and a cyanophilous lichen belonging to the Collemaceae, thriving on the crust at later stages of the succession, are epedaphic (grow above soil surface), and exhibit morphological convergence. The thallus is hard, gelatinous, and resists the impact of rain and wind by swelling rapidly when wetted and preserving the moisture for a prolonged time. Another feature of these two genera, which may be regarded as convergent, is their dark brown colour. This results from the thick, pigmented sheaths, which provide obvious photoprotection in a high-radiation environment (Abeliovich and Shilo 1972). Such a colour is never seen in the endedaphic colonizers *Microcoleus* and *Schizothrix* which emerge from the ground with a blue-green thallus only after being wetted by an effective rain.

Green algae from edaphic algal crusts have been reviewed by Friedmann and Galun (1974), but without specific reference to sandy soils such as stable dunes. Diatoms, coccoid chlorophytes, and xanthophytes of soil crusts are recorded by Johansen (1993). Further research is needed for evaluating the role of green algae and other groups in the desert dune ecosystem. The physiological characteristics of organisms found in the microbiotic crust in the western Negev are outlined by Lange et al. (1990).

5.11.2 Lichens and Fungi

The only lichens that I have observed in the microbiotic crusts developed on stable dunes in the Negev are cyanophilous lichens. Although many terrestrial lichens of desert areas are mentioned in the account by Friedmann and Galun

(1974), these authors do not refer specifically to soil types and therefore one cannot regard their data as being related to sand dunes. A similar poverty of data exists regarding desert fungi. Friedmann and Galun refer to the adaptations of fungi to the desert environment as being mainly the possession of brown-coloured mycelia and spores. This dark colour protects the cells from excessive solar radiation.

5.11.3 Mosses

Mosses occur typically in mesic habitats with high humidities, yet their role in the desert ecosystem is important on dunes. Briefly, their life cycle begins with a germinating spore which gives rise to a branched network of filaments called a protonema. The protonema grows horizontally and can reduce the mobility of sand grains. After developing to an appropriate stage, vertical caulidia start to grow. A caulid resembles a stem with leaf-like structures, and at its base are multicellular, colourless, hair-like filaments known as rhizoids, which function like roots (Schofield 1985).

The caulidia lack the lignified vascular cells and stomata of higher plants. The leaf-like appendages of mosses are usually one cell thick, except for the midrib (costa). They are not true leaves but function as photosynthetic organs. When dry these "leaves" become appressed to the caulidia and they open out when wet. There is a remarkable change in the colour of moss with changes in their moisture status. Several moss species have the ability to produce rhizoids when covered by sand, to elongate, and to grow above the sand cover. Hence, they represent a diminutive version of grasses with the ability to grow vertically above the covering sand. Their "stems" differ from the orthogeocorms of *Stipagrostis scoparia* in this small size, slow rate of growth, and in the ability to become seasonally dormant and air-dry. Being small and growing vertically up to a maximum of 4 cm/year (Moore and Scott 1979) mosses can only grow in relatively stable habitats. In the Negev desert, mosses are able to grow despite constant but slow silt deposition (Danin and Ganor 1991). In more extensive areas they remain part of the microbiotic crust even after the surface has become stable enough to facilitate the existence of epedaphic cyanobacteria and cyanophilous lichens. Mosses may develop in desert crusts where the mean annual rainfall exceeds 90 mm, only if there is sufficient silt and clay in the sandy soil. Observations on plant succession in coastal dunes at Bodega Head, California (Danin et al. 1995) show that *Bryum capillare* is the pioneer colonizer of microbiotic crust on immobile sand, followed by *Didymodon vinealis* and *Brachythecium albicans*.

The reactive growth of mosses on sand is similar to that of grasses. An illustration of *Bryum pendulum*, a coastal moss in Britain (Gimingham 1948; Chapman 1976; Fig. 6.3) resembles in miniature an illustration of a grass belonging to the *Stipagrostis scoparia* group. When covered by sand, tufts of this moss develop leafless continuations of the caulidia which become leafy again above ground. The subterranean sections close to the surface give rise to rhizoids and the old lower parts disintegrate. The formation of rhizoids resembles the formation of nodal roots in perennial grasses. Moore and Scott (1979) studied the mosses of coastal dunes from Victoria, Australia, and recognized several modes of reactive growth. (1) Initiation of growth sites at the apex or near the apex in *Barbula torquata*, *Barbula billardieri*, *Tortula princeps*, and two *Campylopus* species; (2) branching of lateral buds along the buried "stems" in *Tortella calycina*, *Thuidium furfurosum*, and *Triquetrella papillata*; (3) rhizoid and protonema development, enabling regrowth at the new sand surface in *Barbula torquata*; (4) death of the buried plant and regrowth from spores. Moore and Scott (1979) concluded that most species of moss have more than one mode of reactive growth. Moss dominants of the semiarid sandy areas of Australia are *Barbula torquata* and *Tortula princeps*. These are also known from coastal areas (Scott 1982). Mosses do not show reactive growth in response to deflation. Recolonization is by spores. As long as the microbiotic crust exists, the danger of wind erosion is small. Soil grains are tightly glued and entangled together. However, trampling by grazing animals may break the crust and make it vulnerable to wind erosion. A trampled area is portrayed in Fig. 13 and discussed in Section 2.2.4.2.2. Much of the dark area on the Israeli side of the border in Fig. 13 is populated with mosses in the microbiotic crust. Significant climatic change, with a drop in the amount of precipitation or air humidity and a decline in the number of dew nights may also cause a deterioration of the crust.

In the desert dune areas of Israel, mosses grow in two main forms. Some *Bryum* spp., *Aloina* spp., and *Crossidium* spp. form mats that do not project upwards from the flat, stable surface. On the other hand, *Pterygoneurum subsessile* grows in small dense tufts, 5–10 cm in diameter. Sand trapped among its long and narrow leaf tips, almost totally covers the plant. This sand cover prevents abrasion by airborne sand (cf. Sects. 5.10 and 7.3.4). *P. subsessile* forms micronebkas 1–2 cm high.

5.11.4 Summary

Microbiotic crusts are natural components of the sand dune ecosystem of the desert. The crust does not develop in extreme desert areas where mean annual rainfall is lower than 50–80 mm, air humidity is low, and there are too few occasions when dew falls. Cyanobacteria establish themselves first. They fix sand grains, silt and clay. Where moisture conditions permit, mosses, cyanophilous lichens, and probably other organisms start to grow and enrich the organismic diversity of the microbiotic crust community. The organisms inhabiting the microbiotic crust are well adapted to the stabilization of dunes and to stable desert dunes. They are all wind dispersed; their diaspores constantly fall on the ground everywhere. They are poikilohydric and thus can take advantage of even small quantities of water. Filamentous cyanobacteria initially develop below the soil surface and move toward the lighter soil surface when wetted. The cyanobacteria of later seral stages develop on the soil surface, protected from solar radiation by their dark pigments. This component of the microbiotic crust affects the mineral nutrition of the entire ecosystem as a result of the chelating polysaccharides in the cyanobacterial sheaths. Mosses make use of the stabilized sand and the ameliorated moisture regime associated with cyanobacterial activity. They grow slowly above the fine airborne particles which settle among their "leaves" and caulidia. The fungal, green algal, and lichen components of the crust deserve further investigations.

6 Typical Transects in a Few Plant Communities and Their Edaphic Relationships

The previous chapters have been dedicated to the reactive growth of individual species of plants in desert and coastal sand dunes. However, the relationships among species in the same habitat and community may tell us much about their role in the vegetation and the environment. In the following chapter an attempt will be made to expand the temporal and spatial dimensions of Chapter 5. Examples have been selected to represent diverse types of ecosystems.

6.1 A Successional Sequence in the Haluza Dunes, Israel

A study (Danin, unpublished) was conducted in the sandy area of the northwestern Negev (Fig. 71). The sedimentation of the alluvial material and sand which compose the mantle of this area took place in two major steps. The alluvial flood plains of the large wadis draining the Negev Highlands, such as Nahal Nizzana, Nahal Lavan, Nahal Besor, and Nahal Rut, were deposited from ca. 35 000 to 25 000 B.P. (Zilberman 1982, 1989; Goring-Morris and Goldberg 1990). The sediments include mainly silt, clay, and fluviatile sand deposits. Subsequently, during the period 25 000 to 18 000 years B.P., sand was deposited, forming linear dunes and sand sheets (Goring-Morris and Goldberg 1990). According to Rendell et al. (1993) sand was also deposited 6000 to 10 000 years ago. These sands have been subjected to several changes of management and human interference since the 1940s. Some of these changes were documented by Orshan and Zohary (1963), Noy-Meir and Seligman (1979), Danin (1978, 1983, 1987, 1991), Tsoar (1974, 1982), Tsoar and Møller (1986), Thomas and Tsoar (1990), and Goring-Morris and Goldberg (1990). The study area includes a few linear dunes 1–5 km north and west of the highways between the Israeli–Egyptian border near Nizzana in the west, and Ashalim–Revivim–Beer Sheva in the east and north (1–4 in Fig. 71). The entire area was moderately grazed in the 1940s (Noy-Meir and Seligman 1979) and was protected between 1948 and 1967. From 1967 to 1982, during the Israeli occupation of Sinai, Bedouin activity increased in the area, but since 1982 it has ceased. The Bedouin cut lignified plants for fuel, graze domestic animals on both annual and perennial plants, and trample the soil crust. Trampling disrupts the microbiotic crust, opening its surface to wind erosion (Danin 1978, 1987; Danin et al. 1989). The main human activity in the area since 1982 has been

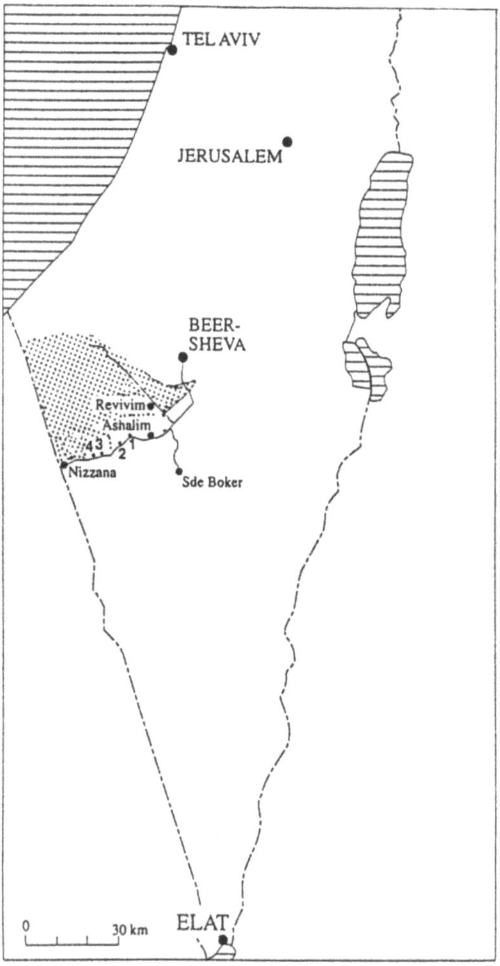

Fig. 71. Map of the study sites of dune succession in the northwestern Negev

localized military maneuvers, vehicle movement, and scattered Bedouin activity. The microbiotic crust is in the process of recovery, leading to the stabilization of formerly mobile sand.

The sites selected for this study were subjectively chosen to represent the most important plant communities and habitats of the sandy area of the western Negev (Fig. 71). They were situated ca. 15–40 km S to SW of Beer Sheva, in several linear dunes between Beer Sheva and Nizzana. The central coordinate was 34°37′E/30°57′N at an elevation of 300–340 m. The area has a mean annual rainfall of 100 mm (Rosenan and Gilead 1985). Vegetation units were distinguished on the dune systems according to species dominance (following Danin et al. 1975). They were graded according to geomorphological and topographical rank from the most mobile to the most stable. Parameters such as ripples, nebkas, “moats”, and the status of the microbiotic crust (Danin 1978, 1991) were used as criteria for evaluating the relative age of sites. In each relevé (a vegetation record), the absolute and relative cover of all the persistent and ephemeral plants was recorded following Danin et al. (1975). The average presence of each species in each community (P) is the number of relevés in which the species occurs, divided by the number of relevés in the community table. The average relative cover (C) of each species in every community was calculated from the sum of relative cover values divided by the number of relevés in the community. The values of the average cover and the average presence of each species in all five communities are given in Table 3. Asso-

Table 3. Synoptic table of communities developed during plant succession on linear dunes, the northern Negev. Numbers of plant communities as in the text

Plant community	6.1.1		6.1.2		6.1.3		6.1.4		6.1.5	
Average cover (persistents)	7.5%		19.3%		22.1%		20.0%		20.6%	
Average cover (ephemerals)	3.5%		14.3%		18.6%		22.3%		16.1%	
Species/releve (avg.)	4.6		16.0		19.7		21.1		23.1	
a. Persistents	C	%P	C	%P	C	%P	C	%P	C	%P
Anabasis articulata			0	9			0	9	0	11
Artemisia monosperma	4	55	48	100	50	100	57	100	10	89
Artactylis carduus			0	36	0	57	0	82	0	89
Calligonum comosum	0	9	1	18			2	27		
Centropodia forsskalii					0	14	0	9		
Convolvulus lanatus			0	9	0	43	0	27	16	89
Cyperus macrorrhizus	0	18	3	82	0	71	1	73	1	78
Echinops philistaeus			3	45	27	100	0	36	1	33
Echiochilon fruticosum									13	56
Haplophyllum tuberculatum			0	9						
Helianthemum sessiliflorum									0	44
Helianthemum stipulatum									0	11
Heliotropium digynum	5	91	3	73	4	43	5	64	5	44
Lycium schweinfurthii					1	14	0	6	1	44
Moltkiopsis ciliata	0	9	9	100	14	100	13	91	13	89
Noaea muronata			0	9			1	18	0	22
Retama raetam			0	9	1	71	3	82	4	89
Stipagrostis ciliata									0	22
Stipagrostis plumosa			0	18	0	71	11	82	30	100
Stipagrostis scoparia	80	100	31	100	1	57	0	45		
Thymelaea hiruta									1	11
b. Ephemerals	C	%P	C	%P	C	%P	C	%P	C	%P
Adonis dentata							0	18	0	11
Ammochloa palaestina							0	9	0	11
Anthemis melampodina					1	57	0	73	0	56
Asphodelus ramosus									0	22
Astragalus annularis			0	9			0	18		

Table 3. *Continued*

b. Ephemerals	C	%P	C	%P	C	%P	C	%P	C	%P
Astragalus caprinus							0	9	0	22
Avena wiestii			0	9			0	9		
Bassia muricata			0	36	0	14	0	9		
Bromus fasciculatus			0	9			0	9		
Carduus getulus			1	64	1	71	0	45	0	22
Colchicum ritchii									0	11
Corynephorus divaricatus			0	9						
Crucianella membranacea					0	14	0	9	0	22
Ctenopsis pectinella					10	45	1	14	4	36
Cutandia memphitica	77	100	45	100	0	43	1	36	0	44
Delphinium peregrinum							0	9		
Eremobium aegyptiacum	23	100	1	55	0	14	0	9		
Erodium crassifolium							0	9	0	22
Erodium laciniatum	0	9	10	91	9	86	14	91	28	100
Erucaria pinnata	0	27	4	91	4	86	5	82	9	89
Filago desertorum							0	9	0	22
Gagea dayana							0	9	0	22
Hippocrepis areolata					0	14	0	36	0	33
Ifloga spicata			5	91	37	100	36	100	12	89
Launaea tenuiloba					0	14	2	27	0	11
Linaria haelava					0	9			0	11
Linaria tenuis							0	9	0	11
Lobularia arabica							0	9	0	11
Lotus halophilus					0	14	0	45	0	22
Matthiola livida			0	18	0	14			0	11
Neurada procumbens							0	18	1	67
Ononis serrata			0	45	0	71	0	36	0	33
Orobanche cernua			0	18	0	43	0	9	0	11
Pancratium sickenbergeri					0	14	0	18	0	78
Papaver humile			0	18						
Paronychia arabica									0	22
Picris asplenioides			0	9	0	43	8	91	21	89
Plantago cylindrica			8	91	6	86	5	91	9	78
Plantago ovata			1	18						
Polycarpon succulentum			0	9						
Pseudorlaya pumila			0	18	0	71	1	55	0	44
Pteranthus dichotomus									0	11
Rumex pictus			5	73	0	100	0	82	0	67
Scabiosa eremophila					0	57			0	33
Schimpera arabica			0	9	0	14	0	18	0	22
Schismus arabicus			0	9						
Senecio glaucus	0	9	10	82	18	100	12	82	7	67
Silene colorata							0	18	0	11
Silene villosa			0	9						
Trifolium tomentosum					0	14			2	9
Trigonella arabica									0	11
Vulpia brevis			0	27	18	86	1	64	4	67

C = average relative percentage of cover, P = percentage of presence in each community ciations are regarded here at the nonranked level of "communities". Five soil profiles, one per community, were sampled at depths of 0–5, 20–25, and 50–55 cm. The texture of these samples was analyzed for the fractions of coarse sand, fine sand, and silt plus clay (Fig. 73).

6.1.1 The *Heliotropium digynum–Stipagrostis scoparia* Community

The relevés for this community were recorded along the vegetated crests of several linear dunes (near points 1, 2, 4 in Fig. 71) which are at present active and mobile. There are large areas with ripples which indicate sand mobility. Many individual annuals and perennials either have exposed roots or are covered with sand. There is no sign of a microbiotic crust. The number of species/relevé in this community is the lowest in the entire study area. Of the two most common persistent species, *Stipagrostis scoparia* is adapted to sand accretion, and *Heliotropium digynum* to sand deflation (Danin 1991; Sect. 5.4.2). Young *S. scoparia* plants occur in flat areas, whereas older ones have mounds of accumulated sand around buried stems. Buried nodes produce new adventitious roots. "Moats" occur near compact and tall mounds (Fig. 8). *Artemisia monosperma*, which displaces *S. scoparia* in the next seral stage, is already present. The two most common annuals which accompany the perennials are *Eremobium aegyptiacum* and *Cutandia memphitica*. Both may be seen in winter and spring, and are resistant both to burial and to deflation. Up to 5–10 cm of their stems may be buried in sand, or 5–10 cm of their root system may be exposed. Adjustment of *Cutandia memphitica* to sand accretion is discussed in Section 5.8.5.

6.1.2 The *Stipagrostis scoparia–Artemisia monosperma* Community

The relevés of this community were recorded in linear dunes which, some 10 years ago, were still mobile. In a few places, younger mobile dunes, populated by the previous plant community (Sect. 6.1.1), exist on the crests above this community (Fig. 72). At present, sand mobility is at low intensity as indicated by the scanty presence of ripples here and there. A crust of cyanobacteria is seen in areas enclosed by grass tufts or by lignified plants. Cyanobacteria are not seen above the ground, but sand grains are already being aggregated by the mucilage on the filaments of the organisms (as in Fig. 10). The remnants of the former stage of the sere are easily recognized by dead or partially dead tufts of *S. scoparia*. Sand mounds in places where *S. scoparia* used to grow are marked by the deteriorating remnants of stems. *Artemisia monosperma* plants often grow on these mounds. Vegetation cover and the number of species are much higher than in the previous community. The decreasing importance of *S. scoparia* (as indicated by its cover and presence in Table 3) and the increasing importance of *A. monosperma* and *Moltkiopsis ciliata*,

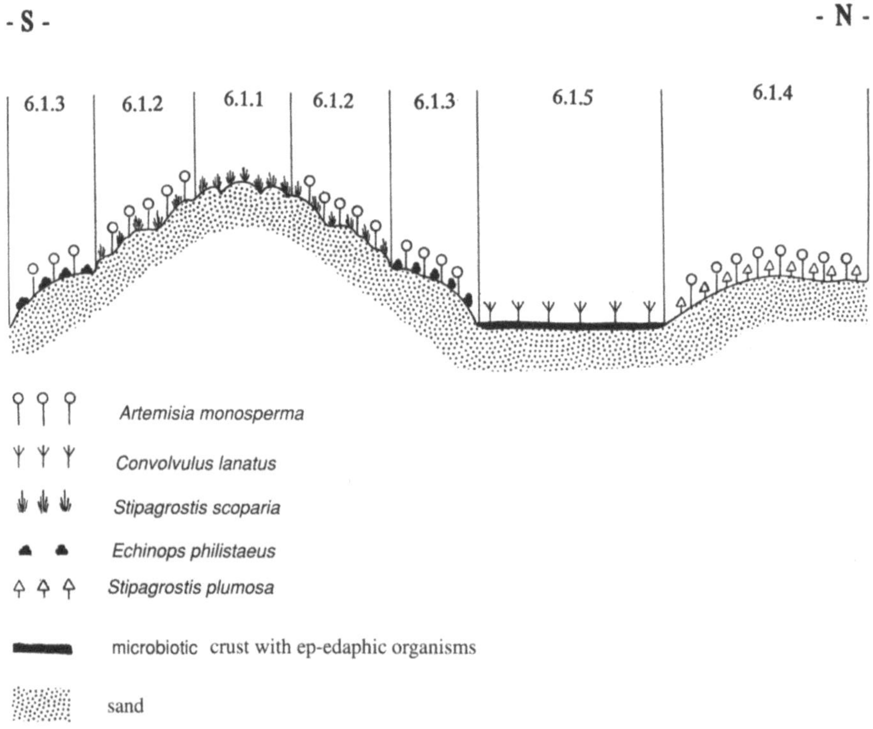

Fig. 72. A schematic presentation of the succession on sand dunes 35 km SW of Beer Sheva. Plant communities numbered according to sections in the text

indicate either increasing sand stability or deflation. The dominant annuals accompanying *Cutandia memphitica* are *Senecio glaucus, Erodium laciniatum*, and *Ctenopsis pectinella.*

6.1.3 The *Echinops philistaeus–Artemisia monosperma* Community

This community was recorded on relatively old parts of the dune (Fig. 72). *Echinops philistaeus* and *Artemisia monosperma* share several properties enabling them to dominate in nearly stable sand. They produce both adventitious roots and new branches from their stems, and their deep lignified tap roots can withstand considerable exposure. Sand deflation is indicated by the importance of *Moltkiopsis ciliata.* The stability and age of the site is indicated by the high values of presence (Table 3) of *Lycium schweinfurthii, Retama raetam*, and *Stipagrostis plumosa. Stipagrostis scoparia, Eremobium aegyptiacum*, and *Cutandia memphitica* – species typical of sites with sand accretion – are found here in small quantities. *Ifloga spicata, Senecio glaucus, Vulpia brevis*, and *Plantago cylindrica* are the dominant annuals of this community.

6.1.4 The *Stipagrostis plumosa–Artemisia monosperma* Community

Most relevés of this community were in sand sheets covering the oldest surfaces. It is not clear whether this community is older or synchronic with that in Section 6.1.3. The most important persistent companions are *Stipagrostis plumosa, Moltkiopsis ciliata*, and *Retama raetam*. These indicate slight mobility of the sand, as well as some deflation or accretion. The dominant annuals of the previous community are accompanied here by *Picris asplenioides* (= *P. radicata*), which is common elsewhere in the Negev on stable sand sheets. There are patches of well-developed microbiotic crust composed of cyanobacteria which grow below the surface.

6.1.5 The *Stipagrostis plumosa–Convolvulus lanatus* Community

The vegetation, landscape features, and extent of development of microbiotic crust indicate a much older terrain than that of the previous communities. Several species adapted to sand deflation have high values of presence and cover in this community (Table 3). They include *Artemisia monosperma, Convolvulus lanatus, Echiochilon fruticosum, Heliotropium digynum*, and *Moltkiopsis ciliata*. Important annuals are *Erodium laciniatum, Erucaria pinnata* (= *E. uncata*), *Ifloga spicata, Picris asplenioides*, and *Plantago cylindrica*.

The microbiotic crust is well developed and is composed of cyanobacteria with considerable biomass above the surface. Their pigments give a dark colour to the soil surface. There are many patches of mosses in the crust as well. The continuous crust cover is broken by the activity of burrowing animals which bring up sand from beneath the crust. The dominant plant on the piles of soil at the sites of burrows is *Vulpia brevis*.

6.1.6 Changes in Soil Texture During Plant Succession

Soil profiles of the five communities discussed above (Sects. 6.1.1 through 6.1.5) are displayed in Fig. 73 and are numbered respectively. There is a general trend, during succession, for an increasing proportion of fine-grained particles to accumulate (Fig. 73). Except for its deeper soil layers (C), the texture of profile 6.1.4 is rather similar to that of 6.1.2. In none of the stages represented by samples 6.1.1 to 6.1.4, does silt plus clay play any important role in soil composition, but in 6.1.5 clay is prominent at the soil surface. The continuous microbiotic crust, mentioned in Sections 2.2.4 and 5.11, seems to be associated with this increase in the silt plus clay content (Danin et al. 1989; Danin 1991).

6.1.7 Summary

The processes of succession discussed above involve changes both in soil and in vegetation. The changes in the main attributes, life histories, and adaptive

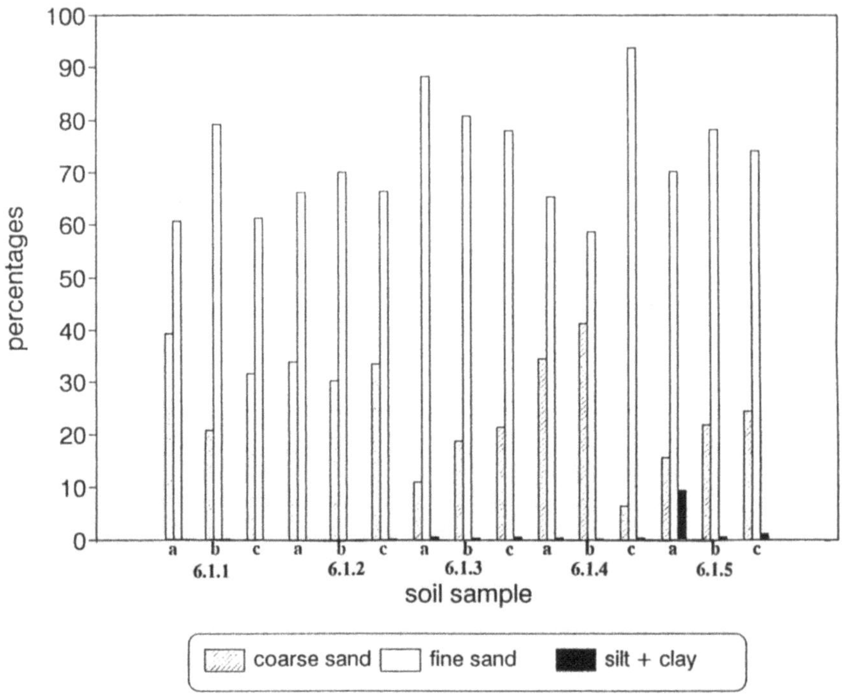

Fig. 73. Soil texture analyses of the seral plant communities (numbered according to relevant sections in the text)

syndromes of the plants in the various communities display the state of mobility they have reached. The dominance of species best adapted to sand accretion, *Stipagrostis scoparia* and *Eremobium aegyptiacum* in the community in Section 6.1.1, decreases to zero in the community in Section 6.1.5. Plants adapted to stable sand, such as *Stipagrostis ciliata*, show an opposite trend in occurrence.

During succession there are also changes in soil texture. The increasing density of plants decreases the wind velocity near the ground and induces the increasing deposition of fine-grained particles. The most advanced stage, the community in Section 6.1.5, has a high percentage of silt plus clay. The amelioration of edaphic conditions is demonstrated by the increasing average cover of persistent species and by the number of species per relevé.

6.2 Stable Dunes on Old Sandstones in the Yamin–Rotem Plain, Israel

The undulating plain of the Yamin–Rotem syncline was studied by Danin et al. (1964), and I made additional observations in the same area in Spring 1994.

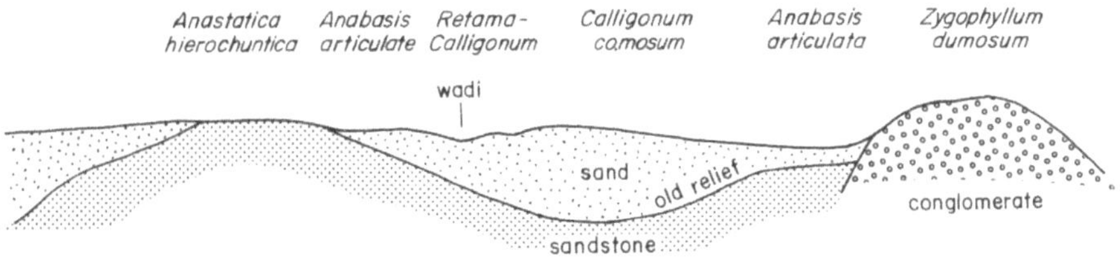

Fig. 74. Relationships of plant communities and soils in Mishor Rotem-Yamin, the northern Negev

Neogene sandstones, conglomerates, and clays once filled this basin to a depth of 180 m. In time, this material has weathered, and loose sand has been deposited over the ancient rocks (Fig. 74). Sites where the conglomerates are exposed are covered with a community dominated by *Zygophyllum dumosum*. Sites where sandstone is exposed are nearly devoid of semishrubs but may support the peculiar lignified annual *Anastatica hierochuntica* (Danin 1983). In the wadis, large shrubs of *Calligonum comosum* and *Retama raetam* grow together with many members of the other communities which occur in the catchment area. The two communities in which semishrubs grow on a sand sheet and show prominent adaptations to dunes are dominated by *Anabasis articulata* on shallow sand, and by *Calligonum comosum* on deep sand. The entire area has been fenced off and protected from grazing and trampling for more than 30 years. There is a well-developed microbiotic crust on the sand surface, the main components of which are filamentous cyanobacteria and patches of *Pterygoneurum*, a moss which tends to form micronebkas. This development of the crust may be related to the moisture regime of 100 mm mean annual rainfall and some 200 nights with high air humidity (Danin 1986). There are small nebkas around the *Anabasis articulata* plants (which do not form shootborne roots). There are also large areas of exposed, deflated ground. A common companion species is *Stipagrostis obtusa* (Sect. 5.6.1) which forms circles with an empty center indicating negligible sand accretion. Many tufts of *S. obtusa* are situated higher than their surroundings or even with depressions nearby (as in Figs. 4, 8). Plant circles are also formed by *Stipagrostis ciliata* (Fig. 56) and by *Asphodelus ramosus* (Figs. 58, 59). These indicate at least a few hundred years of stability (Sect. 5.6.4).

The area dominated by *Calligonum comosum* has an undulating topography with prominent nebkas around and within the dominant plants. These nebkas sometimes form circles or elliptic mounds, 5–11 m in diameter and 1–1.5 m higher than their surroundings. The most consistent species of *Stipagrostis* in this community is *S. plumosa*, and there are scarcely any individuals of *S. obtusa* or *S. ciliata* there. *S. plumosa* indicates moderate sand accretion or stable sand (Sect. 5.3.1). Circles of *Asphodelus ramosus* are present too, indicating long stability in the area. The dominance of *Calligonum comosum* and the high frequency of *Stipagrostis plumosa* implies that this vegetation and landscape unit experienced a high rate of sand accretion in the past. However, circles of *Asphodelus ramosus* and the presence of a microbiotic crust indicate stability at the present time. The entire area may be regarded as being stable now, but stability in the *Anabasis articulata*-dominated patches is greater, and has lasted for a longer time, than in the *Calligonum comosum* patches. The possibility that constant grazing and trampling pressures existed before protection was provided by the fence, more than 30 years ago, should be examined. If such pressure did take place, it could have had an impact on sand mobility.

To conclude, the ecomorphological syndromes serve well as indicators of sand mobility. There is not even one plant in the group that requires sand accumulation (Sect. 5.1). The area with deep sand is dominated by a species which can stand intensive sand mobility as well as complete stability. Each companion perennial grass displays high fidelity to the plant community which is closest in environmental conditions to its syndrome. The microbiotic crust further indicates sand stability and low artificial destruction by trampling etc.

6.3 Dunes Covering Stable Sand Sheets in Northern Sinai

Linear sand dunes in northwestern Sinai, near Tasa, look similar to the dunes in Fig. 3. Fine-grained sand covers an area which was once mobile, later became stabilized, and is now exposed to wind erosion. Erosion is clearly indicated by the exposed tap roots of the prevailing plants – *Convolvulus lanatus, Cornulaca monacantha*, and *Artemisia monosperma* (Figs. 49, 50, 52). In some places, roots may be exposed for 1–2 m. There are large sand particles, grit, and even pebbles at the soil surface, indicating wind removal of small particles and enrichment of the surface by large particles. Vegetation cover is low (<5%) and the plants are small. Dune crests are devoid of plants, whereas the dunes slopes are vegetated. The upper slopes are dominated by *Stipagrostis scoparia*, a plant resistant to sand accretion (Sect. 5.1.1). The lower slopes, where the area of sand accretion meets the area of sand deflation, are covered by large and dense shrubs such as *Artemisia monosperma* and *Scrophularia hypericifolia*. Plant density, plant size, and the origin of plants indicate better environmental conditions than in the surrounding area. This position is relatively protected from erosion. Plant roots do not become exposed as in the open interdune area. Instead, they are constantly being covered by sand and have the ability to produce adventitious roots. The large size and high density they achieve may be interpreted as having access to larger amounts of water than in the interdune area. The additional water supply is assumed to be below ground, charged by vertical water movement in dune areas as reported by Prill (1968). Apart from deep root systems and the ability to produce adventitious roots, plants growing here show no special adaptations.

6.4 Sand Covering Salt Marshes in Sinai, Namibia, and California

Salt marshes are a common feature of coastal and inland depression areas in deserts (Danin, 1981a). Many hydrohalophytes or phreatophytic halophytes trap sand and form nebkas (Fig. 16; Sect. 2.2.3; Batanouny and Batanouny 1968, 1969; Chapman 1976; Danin 1983; Evenari et al. 1985; Bendali 1987). According to Batanouny and Batanouny (1968, 1969) the plants produce adventitious roots, and new branches sprout into the accumulated sand so that the plants continually "float" to the surface. In most cases studied by Evenari et al. (1985), the plants do not drive adventitious roots into the nebka. The nebka-producing halophytes studied by Evenari et al. (1985) were *Nitraria retusa*, *Zygophyllum album*, *Limonium axillare*, *Aeluropus lagopoides*, and *Halocnemum strobilaceum*.

Huge areas of coastal wet saline soil in Namibia, south of Walvis Bay, are covered by an almost monospecific community of *Salsola nollothensis*, the plants of which form nebkas more than 2 m high. A similar landscape exists in the Tarim Basin of Central Asia, with *Nitraria schoberi* forming large nebkas in an area with brackish underground water (Walter and Box 1983b).

Many of the desert sand-dune or fixed sand areas of Owens Valley and Death Valley in California cover salt marshes or alkali sinks. These areas support hydrohalophytes which tend to form monospecific communities. A sand dune near Keeler, in Owens Valley, is covered by large nebkas of *Sarcobatus vermiculatus*. This plant produces shootborne roots which help build relatively large nebkas. There are no successional changes in the nebka communities dominated by *Halocnemum strobilaceum* in northern Sinai, by *S. nollothensis* in Namibia, or by *Sarcobatus vermiculatus* in California.

The hypothesis of Batanouny and Batanouny (1968, 1969) that the ability to produce phytogenic hillocks is an adaptation to desert dunes, seems to be only partially true. Many hydrohalophytes do not produce roots in the accumulated sand, in contrast with phytogenic hillocks in nonsaline habitats. Edaphic conditions in sandy areas, with *Stipagrostis scoparia*, *Artemisia monosperma*, or other sand-fixing plants, change as a result of plant activity; but salt marsh vegetation, such as that illustrated in Fig. 16, will not alter as a result of small nebkas accumulated around the hydrohalophytes. The principal limiting factor of this habitat, the occurrence of saline water close to the surface, does not change. The principal adaptation of halophytes is to withstand a wet and saline rhizosphere and not to trap sand; the latter process may be regarded as a by-product and not as a main adaptation.

6.5 Sand Covering Fresh Water Aquifers

West of the coastal salt marshes in southeastern Sinai the glycophytic tree *Salvadora persica* produces nebkas more than 5 m high (Fig. 75; Danin 1983;

Fig. 75. *Salvadora persica*, a phreatophyte, with a large nebka in a sandy alluvial fan near Nabq, SE coastal area of Sinai. Trunks are exposed on the windward, northern, side of the nebka

Fig. 76. Three date palms planted near a salt marsh in northern Sinai and partially covered by a sand dune

Evenari et al. 1985). It is assumed that its roots reach nonsaline water. In a similar way, date palms (*Phoenix dactylifera*) planted near salt marshes may become substantially covered with sand while still being able to use a fresh-water table (Fig. 76).

Sand may move and cover sites with a high water table of fresh water. These sites will eventually become appropriate for the establishment of phreatophytes which may then prevail for a long time. Such a plant community, dominated by *Artemisia monosperma*, exists in the Negev a few kilometers west of Revivim. *A. monosperma* is accompanied by *Prosopis farcta*, but this is restricted to places with a high water table. Even in the nondesert areas of Israel *P. farcta* does not establish itself from seeds but multiplies vegetatively (Dafni 1975; Feinbrun-Dothan and Danin 1991). The desert sites in which *P. farcta* is found today are old terraces of wadis, some of which have fresh-water springs downstream, further west. Some time in the past several individuals must have germinated in a few places and established their roots in the fresh-water aquifer, enabling them now to grow in association with psammophytes.

Prosopis glandulosa is a common phreatophyte of the southwestern USA and Mexico (Munz and Keck 1963; Hickman 1993). It may use fresh or slightly alkaline aquifers. In Death Valley, California, there is a large area of sand dunes where much of the sand is in very large nebkas which have accumulated around large trees or shrubs of *P. glandulosa*. The plants have adventitious roots emerging from thick lignified branches. Surrounded by many salt marsh plants, it resembles the situation described above for *Salvadora persica*.

Tamarix aphylla plays a similar role in the Arava Valley and in southern Sinai where it typically forms large nebkas in sandy areas (Danin 1983, p. 62). There are many specimens of *Tamarix aphylla* in depressions and on sand dunes in the desert sands east of El Arish and west of the Israeli–Egyptian border. Several trees occur in rows along or close to the depressions among the dunes. Such are the five trees at the lower left corner of Fig. 77. These trees may have been planted to decrease sand mobility, but their sparse and inconsistent occurrence indicates that they might have become established during rare events of rising fresh-water table in depressions among the dunes. Wet sites with no competitors are the appropriate habitats for germination and establishment of this tree (Danin 1981b).

6.6 Sand Covering Limestone Hills, Western Sinai

The study area, some 20 km NE of Suez, west of the Mitla Pass, near Wadi el Haj, has a series of habitats typified by different substrates, quantity of sand, depth of the sand layer, and degrees of sand mobility (Fig. 78). The depth is influenced by the topographic position and the direction of strong winds,

Fig. 77. Aerial view of the coastal dunes east of El-'Arish, Sinai. The large round trees on the dunes, sometimes in rows of 3–5 trees, are *Tamarix aphylla*. Possibly the trees germinated and became established when a fresh-water table was close to the surface for a few weeks during the summer

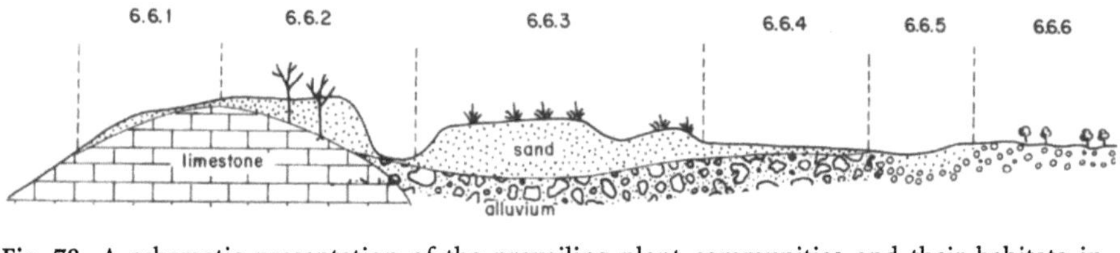

Fig. 78. A schematic presentation of the prevailing plant communities and their habitats in western Sinai, ca. 20 km NE of Suez. Plant communities numbered according to relevant sections in the text

capable of moving sand. The vegetation in each area reflects the hazards related to sand mobility and the adaptation of plants to these hazards.

6.6.1 *Anabasis articulata–Astragalus camelorum* Community on Shallow Sand

This habitat is restricted to areas where the amount of mobile sand is not sufficient to create a deep mantle over the hills; sand depth is 10–70 cm. This habitat supplies larger quantities of water than pure sand because of the high water-holding capacity of the underlying lithosol, which is rich in fine-grained particles (Danin 1983), and because of the mulching effect of the sand cover (Fig. 78). Plants growing here have to survive constant covering and removal of sand from their aboveground parts. A conspicuous companion to the dominant *Anabasis articulata*, is the rare semishrub *Astragalus camelorum* (Danin 1976). It was regarded as an endemic to western and northwestern Sinai for many years but was found recently in the Arava Valley, in sandy ground, by Baierle (1993). *Astragalus camelorum* is a stem-assimilant, totally covered by appressed hairs, which sheds its leaves while the fruits are ripening and then remains leafless for a long time. The diaspore is a short and rounded legume, densely covered by long hairs, resembling the rolling diaspores of *Calligonum* species.

6.6.2 *Retama raetam* Shrubs on Sands up to 2 m Deep

Sites where sand depth is 1–2 m are populated with stands of *Retama raetam* (Fig. 79). The most common habitats of this species in the Sinai desert are wadis where additional water is below ground. In the sands of this habitat (Fig. 78, number 6.6.2), *R. raetam* grows in a diffused pattern (i.e., it is not restricted to wadis; mode diffuse, Monod 1931) of open, monospecific stands. Its crown may be free of or nearly covered by sand. Total burial causes death, but plants may survive cycles of accumulation and deflation down to the lithosol, where

Fig. 79. *Retama raetam* the only plant growing on limestone hills in western Sinai, is covered from time to time by 1–2 m of mobile sand

the principal rhizosphere of the plant exists. No adventitious roots occur on buried shoots of *R. raetam* (Sect. 5.5.3).

6.6.3 *Stipagrostis scoparia* Dominating Deep Mobile Sand

Where sand depth exceeds 3 m, vegetation characteristic of mobile sand develops with *Stipagrostis scoparia* and *S. acutiflora*. These two species are so well-adapted to sand mobility that they require it for survival (see Sects. 5.1.1 and 5.1.2).

6.6.4 *Anabasis articulata–Fagonia glutinosa* Community on Sand Covering Alluvial Plain

As on the stony hill slope (Sect. 6.6.1) the dominant plant is *Anabasis articulata*. However, it grows here at a lower density, there is no *Astragalus camelorum*, and its companions also differ. The rainy winter of 1975 was accompanied by massive germination and establishment of *Fagonia glutinosa*, the leaves of which are typically covered by sand fixed to its viscid surface (Sect. 5.10.4; Fig. 70). Many individuals were annuals but a few became lignified and perennial.

6.6.5 *Anabasis articulata–Artemisia monosperma* Community

The soils of wadis draining the alluvial fans in this area are sandy, and the dominant plant is *Anabasis articulata* which covers considerable areas of extreme desert in Sinai and the southern Negev (Danin 1983). Its companion, *Artemisia monosperma* (Sect. 5.2.2) reduces wind speed locally and enhances sand deposition in the wadi beds.

6.6.6 Reg Vegetation in Wadis with Nonsandy Substrate

The gravel plain has but a small amount of sand in its soil. There are many plant communities in these wadis, depending on the order of the wadi and on the specific soil conditions of the catchment area and the wadi beds. For more information on this complex mosaic of vegetation and soil see Danin (1983).

6.6.7 Summary

The transect in W Sinai has implications for investigators of vegetation in other parts of the world. The rare *Astragalus camelorum* is confined to sites of lithosol covered by a layer of sand for a limited time and to a limited depth. A thicker sand layer typifies the monospecific community of *Retama raetam* which is covered by sand when the wind blows in a certain direction. Deep and mobile dunes support two species of *Stipagrostis* (Sect. 5.1) which grow only in this habitat. At the boundary of the extreme desert, fine-grained soils support vegetation only in wadis whereas a transitional texture supports *Fagonia glutinosa* which may be an annual in relatively dry microhabitats or perennial in moister regions.

6.7 Eureka Dunes, California

Eureka Dunes are situated at the southern end of Eureka Valley, Inyo County, California. They form an oblong ridge, 5 km long, about 1.6 km wide, and 230 m high (DeDecker 1976). They are surrounded by alluvial fans on the north, a playa to the west, and the rocky Last Chance Mountains to the south and east. Their flora and phytogeographical relationships were studied by DeDecker (1976) and Pavlik (1985). The vegetation at the southern part of the valley coincides with the geomorphological-edaphic units illustrated in Fig. 80 and discussed below.

6.7.1 Alluvial Fans Dominated by *Larrea tridentata*

Most of Eureka Valley consists of rather flat alluvial fans, and stones covered with dark rock varnish. Vegetation is restricted to wadis and dominated by the shrub *Larrea tridentata.* According to DeDecker (1976), *Larrea* is accompanied by carpets of showy, colourful herbaceous annuals in rainy years. Although the alluvium contains fine- and coarse-grained particles among the stones, it is not enough to support typical sand vegetation.

6.7.2 Sand Covering Alluvium, Dominated by *Atriplex polycarpa*

Where a sand layer a few centimeters thick covers the alluvium (and probably the playa) at the periphery of Eureka Dunes, the vegetation is an open

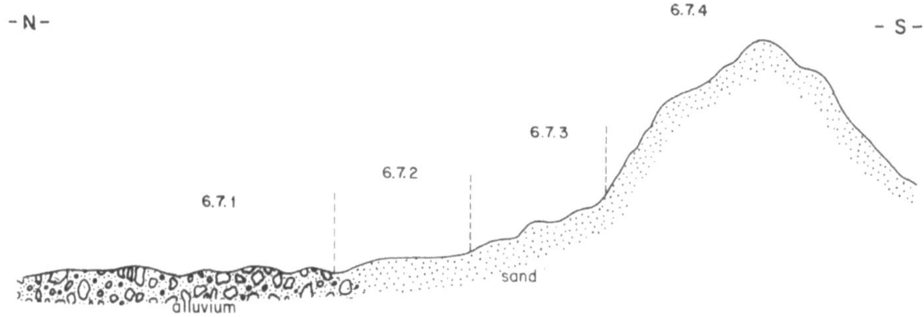

Fig. 80. A schematic presentation of the prevailing plant communities and their habitats in Eureka Dunes, California. Plant communities numbered according to relevant sections in the text

monospecific community of *Atriplex polycarpa* (DeDecker 1976; see also Sect. 6.9.3). The impact of sand covering finer soil in desert results in similar vegetation in other regions. For example where sand covers limestone slopes in Biqa'at Uvda, Israel, several semishrub xerohalophytes, such as *Salsola tetrandra*, *Agathophora alopecuroides*, and *Traganum nudatum* develop in a diffuse pattern, accompanied by many annual psammophytes (Danin, unpubl. data). In both cases the sand layer increases the amount of water penetrating the soil to a depth where it is protected from direct evaporation (see Sect. 3.2.2).

6.7.3 Thick and Slightly Mobile Sand Dominated by *Psorothamnus polyadenius*

The transition from the *Atriplex*-dominated area to the lower part of the dune proper (plinth) is sharp. Here sand is so deep that vegetation appears to be influenced only by the edaphic conditions of the sand and not by the underlying substrate. This unit may be characterized by nebkas formed by the dominant shrub *Psorothamnus polyadenius* (Sect. 5.2.7) accompanied by many other plant species. Sand mobility is not as high here as in the following zone (Sect. 6.7.4), as indicated by the presence of *Achnatherum hymenoides* (Sect. 5.6.3) which grows in dense stands where sand is immobile. Among the many plants listed by DeDecker (1976) from this zone is the endemic *Oenothera ovita* subsp. *eurekensis* (= *O. californica* subsp. *eurekensis*) which withstands sand by producing adventitious roots and growing above the sand cover. *P. polyadenius* behaves similarly. The adaptations of *Tiquilia plicata* and *Dicoria canescens* are discussed in detail in Sections 5.3.7 and 5.10.6, respectively. Both species achieve high density among the shrubs in certain years.

6.7.4 Mobile Sand Dominated by *Swallenia alexandrae*

The dominant, and almost the only, perennial plant growing in the main zone of mobile sand is the narrow endemic *Swallenia alexandrae*. Some plants form nebkas, indicating sand accretion. At the end of the summer these are the most viable and green individuals. Other plants have exposed vertical stems up to 50 cm high. These specimens are still green. Scars of nodal roots found on old stems indicate their past function as orthogeocorms. Such long stems probably gave DeDecker (1976) the mistaken impression that the plant reproduces from rhizomes. The *Swallenia* tufts are sparse and the constant annual changes in wind direction lead to a random cover or exposure of plants. The species does not grow extensively enough to stabilize the sand nor to contribute to succession (as discussed for *Stipagrostis scoparia* in Sect. 6.1).

6.7.5 Summary

The transect of Eureka Dunes shares similarities with the transect of W Sinai (Sect. 6.6) in the following ways: (1) the area surrounding the dunes and sandy soils supports vegetation that is restricted to wadis; (2) a diffuse community occurs on shallow sand cover; and (3) plants which require sand accretion for survival have orthogeocorms.

6.8 The Kelso Dunes, California

The Kelso Dunes are the largest area of desert dunes in the southwestern USA (Lancaster 1993). They are situated in the eastern Mojave Desert, some 280 km NE of Los Angeles at coordinates 115°42′W/34°54′N. Their long Pleistocene-to-present history (Lancaster 1993) has produced a complicated mosaic of dunes with varying texture, form, degree of stability, and vegetation. The transect discussed here was studied in December 1994 at the southeastern section of the Kelso Dunes (for a satellite image of the area see Lancaster 1993; Fig. 3). The relations of the vegetation and its edaphic units are schematically presented in Fig. 81. The dunes constitute an oblong ridge, nearly 7 km long, 2.7 km wide, and up to 200 m high, running in a west-southwest–east-northeast direction. They are partly bounded on their southern and eastern sides by alluvial fans, and on their northern and western sides by stable sands of various ages. The sand is derived from rocks weathered in the Mojave Desert and transported by the Mojave River, as discussed in detail by Lancaster (1993). The vegetation found at the south-eastern part of the dune coincides with geomorphological-edaphic units as illustrated in Fig. 81 and discussed below.

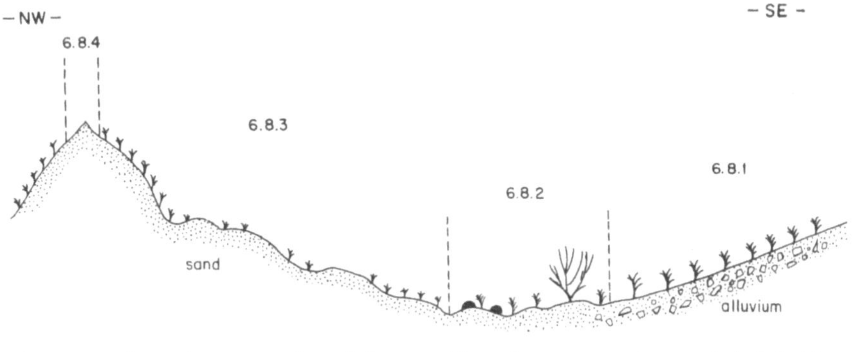

Fig. 81. A schematic presentation of the prevailing plant communities and their habitats in the Kelso Dunes, California. Plant communities numbered according to relevant sections in the text

6.8.1 Alluvial Plain Dominated by *Larrea tridentata*

The sandy alluvial fans at the foot of Granite Mountains support a shrub steppe dominated by *Larrea tridentata* and *Ambrosia dumosa* – a vegetation type typical of the gentle slopes (bajadas) and valleys of the Mojave desert (Fonteyn and Mahall 1978). The *L. tridentata* bushes are 0.5–1.5 m tall and many of them display a circular pattern. As discussed by Barbour (1969) and Vasek (1980), the circular pattern of *L. tridentata* develops over a long period of stability of the community. The main disturbances at the site at present are grazing and trampling by cattle and some trampling by visitors. A crust of filamentous cyanobacteria and cyanophilous lichens is present in some places, such as below the canopy of *Larrea* shrubs, which are relatively protected from trampling. From comparison with other areas, I assume that if the trampling factor were removed, the entire intershrub area could be covered by such a crust. A prominent companion is a species of *Opuntia.*

6.8.2 Plinth Dominated by *Hilaria rigida*

The transition of vegetation from the alluvium to the sand is very sharp. The *Opuntia* and most of the shrub companions of *Larrea tridentata* disappear, whereas grasses adapted to sand appear. The dominant grass of the plinth is *Hilaria rigida* which forms nebkas. In late summer and fall, most *H. rigida* plants still have green culms, at least in their lower internodes. There are many tufts of *Achnatherum hymenoides* scattered throughout the area. Both grasses indicates that there is some sand accretion, but the extent of it is low. The most prominent indicators of the plinth are large *Larrea tridentata* shrubs, 3–4 m tall, most of which have nebkas. Where the west-facing part of the nebka is eroded, these display many adventitious roots which originally developed from buried branches. Closer to the main sand body of the Kelso Dunes there are no *Larrea tridentata* shrubs and sand mobility seems to be greater, as evidenced by higher nebkas and by an increasing number of plants with exposed roots. *Croton californicus*, which has silvery-white leaves, is a prominent companion of *H. rigida.* The white leaves are densely covered by appressed stellate multicellular trichomes, to such an extent that hardly any green part of the leaf is visible. It has an open canopy which is not an obstacle to sand movement. *C. californicus* resists low rates of sand accretion and sand deflation and forms small nebkas, but it does not produce shootborne roots. A species which builds larger nebkas is *Petalonyx thurberi*, which has conspicuous corky bark on its stems. Such a corky layer is regarded by Fahn (1982) as a measure of protection from desiccation; it could also be interpreted as providing protection from abrasion by airborne sand. I did not find any adventitious roots emerging from the buried branches. *Hilaria*, *Croton*, and *Petalonyx thurberi* are the main contributors to the vegetation cover. Closer to the boundary with the next unit are a few patches of *Panicum urvilleanum* and areas of mobile sand with no plants growing in them.

6.8.3 Mobile Sand Dominated by *Panicum urvilleanum*

The dominant plant of the main sand body of the active dunes is *Panicum urvilleanum*. It grows in small patches or single occasional tufts on the mobile crests, but creates continuous patches on lower slopes where the sand has become relatively stable. *Panicum urvilleanum* has a network of rhizodes from which orthogeocorms develop. The whole south-facing slope of the Kelso Dunes is composed of small crescentic dunes, 3–8 m high, 50–150 m apart, and nearly devoid of plants. Among these crests are lower, plant-covered and stabilized dunes with *Achnatherum hymenoides* and *Hilaria rigida*. Occasional well-developed trees of *Chilopsis linearis* are found close to the steep higher slopes of the dune. Their occurrence here is hard to explain as they are mainly riparian plants (Hickman 1993). The steep slopes are densely covered by *P. urvilleanum*. Paths indicate trampling by grazing animals.

6.8.4 Dune Crest

The dune crest is devoid of vegetation, as illustrated in Lancaster (1993; Figs. 1 and 7). The intense, constant mobility of sand prevents the establishment of any plants, a common phenomenon on all seif dunes.

6.8.5 Summary

The transect in the Kelso Dunes resembles that of Eureka Dunes (Sect. 6.7) in several ways. The alluvial plains of both transects are dominated by *Larrea tridentata*, although its companions differ as a result of edaphic, climatic, and floristic differences between the locations. The plinth vegetation of the Kelso Dunes is rich in perennial grasses, whereas that of the Eureka Dunes has a different assemblage of plants adapted to low sand mobility. The main area of mobile sand, however, is similarly covered in both cases by a tough perennial grass which has orthogeocorms.

7 Plant Adaptations to Environmental Stresses in Desert Dunes

In this chapter I discuss the adaptations of higher plants to desert dunes as a consequence of the case histories and typical transects presented in the previous chapters.

7.1 Convergence and Adaptation

Convergence is known (Lapedes 1976) as the "development of similarities between animals or plants of different groups resulting from adaptation to similar habitats". A common example of convergence is the "cactoid" form of plants from the American Cactaceae and some African species of Euphorbiaceae (Rost et al. 1984; Niklas 1992). Rost et al. wrote "Similar traits sometimes arise independently in quite unrelated groups of plants. Convergence often occurs when plants of different origins evolve under similar environmental pressure". The present book, in its study of morphological adaptations to the dune environment, is an attempt to demonstrate convergence in different floristic regions. Species from different sections of a genus or genera, which display a similar syndrome of morphological adaptations, testify to the existence of convergence. The common features of groups of species confined to specific dune habitats from widely disjunct locations prove the hypothesis that some traits represent major adaptations to specific dune stresses.

7.2 Reactive Growth and Sand Mobility

The instability, or mobility, of the ground in sand dunes seems to be the most destructive and hence the most significant environmental pressure exerted by the dunes. The ability of plants to respond immediately to the mobility of the soil affects their survival. For many years the concept of "mobility" among plants was mainly applied to seed dispersal (Weaver and Clements 1929). While dealing with the architecture of tropical plants, Halle et al. (1978) used the term "vegetative mobility" to describe plants with stolons that can move and expand laterally. They also called "mobile" those epiphytes which die back basally so that the plant gradually moves upward. Mosses behave similarly in desert crusts. Danin and Ganor (1991) found that only the upper 1–2 mm of a moss cushion is alive, whereas remnants of dead mosses could be detected down to a depth of more than 10 mm in silty soil. Similarly, dune plants which are covered by sand have to move upward. Plants with exposed roots have to be able to produce new shoots and move downward. Due to the difficulties in the germination and establishment of small seedlings in an unpredictable, dry, and unstable habitat, plants possessing "vegetative mobility" have a biological advantage over others in mobile sand. The mobility of dune plants was regarded in previous chapters as "reactive growth".

7.2.1 Adjustment to Changing Soil Surface

The instability of sand dunes and the possibilities of erosion or deflation of the sand surface, even after initial stabilization, put additional stress on plants. Plants may either be indifferent to, or be able to, respond to changes in the soil surface by moving up or down.

7.2.1.1 Species Indifferent to Changes in the Soil Surface

The establishment of the plants in this group is not of concern here. Their reaction to sand covering their crowns or exposure of their roots as a result of sand accretion is discussed in Sections 5.5 and 5.7. *Retama raetam* may remain alive and function normally when its roots are exposed (Fig. 50) or when its crown is covered by sand (Fig. 79). In a similar way, *Haloxylon persicum* can tolerate both processes (Fig. 63). Adventitious roots or stems are not formed. For example, in the coastal strand of Israel, plants of *Thymelaea hirsuta* growing in crevices of sandstone may to a large extent be covered by sand without developing of adventitious roots.

7.2.1.2 Species Adjusting the Crown Position of Their Seedlings

Germination is a crucial step in plant life. When diaspores of desert plants are exposed to the dry air, most do not germinate. However, if covered by sand, they begin to germinate, and the seedlings tolerate being covered by varying amounts of sand. The first kind of reactive growth of the seedlings of *Cutandia memphitica* is elongation of the mesocotyl. This enables the leaves of the crown to be placed at the right position. Additional adjustments are made by elongation of leaf sheaths and first internodes (Sects. 5.8.5 and 6.1.1). A similar type of reactive growth is exhibited in *Dicoria canescens*; the hypocotyl pushes the apical meristem of the seedling's stem to a position above the ground (Sects. 5.10.6 and 6.7.3). Along the Californian coast, seedlings of the strand annual *Cakile maritima* typically have their cotyledons on the sand surface but germination occurs at depths of 3–10 cm (Barbour et al. 1973). Thus, the emergence of the cotyledons and the apical meristem is carried out by elongation of the hypocotyl. I expect that these mechanisms of emergence are common to seedlings of many other dune plants.

7.2.1.3 Renewal Buds Close to Soil Surface

Hemicryptophytes and geophytes of relatively stable sand sheets must be preadapted to the hazards of being covered or exposed as a result of local changes in the level of soil surface. The plants described in previous chapters display two main syndromes, those with rhizomes and those with rhizodes. *Asphodelus ramosus* adjusts to rapid sand accretion by producing a vertical leafy rhizome (Figs. 60, 61) and to deflation by contraction of its roots (Sect. 5.6.4; Fig. 57). The plasticity of *Cyperus macrorrhizus* to sand accretion is achieved by reactive elongation of otherwise short internodes (Figs. 35, 36, 37, 39), and to sand deflation by sprouting buds whose dormancy is removed when exposed (Fig. 38).

Plants with rhizodes may react to sand accretion by a slight change in the angle of their usually horizontal rhizodes. Examples are *Panicum turgidum* (Fig. 30), *Cyperus capitatus* (Fig. 67), and *Cyperus sharonensis* (Fig. 68).

7.2.1.4 Species With Renewal Buds Above Soil Surface

Shrubs and trees adapted to being covered by sand produce adventitious roots from their covered stems (Fig. 29); in time new branches sprout from above the roots. Thus, the entire plant moves functionally upwards as old branches die and disintegrate (Fig. 24d). Many shrubs which respond in this way, as well as trees such as *Tamarix aphylla* and *Prosopis glandulosa*, are described in Section 5.2

The grass species which require sand are capable of responding to sand accretion in a similar way to shrubs and trees. However, they may do it faster. Roots of *Stipagrostis pungens* (Bendali 1987; Bendali et al. 1990) grow at a rate of 1 cm/day. These grasses are more vigorous in sites of sand accretion than on stable dunes. This was proved experimentally for *Ammophila breviligulata* (Disraeli 1984; Seliskar 1994). The capacity for leaf sheath elongation as a mechanism for placing the leaf blade above ground (e.g. *Panicum urvilleanum*, Sect. 5.1.7), is likely to be a common feature of many grasses confined to this type of habitat. The species discussed in Section 5.3 function in a similar way, but the extent or speed of their reactive growth may be smaller or slower.

7.2.1.5 Rootborne Shoots

By producing shoots from their exposed roots, several species of the Boraginaceae (Sect. 5.4) are able to follow the deflating soil surface. They have horizontal succulent roots at varying depths from which new shoots sprout when exposed. Thus the vegetatively produced offspring are adjusted to the new soil surface even when the mother plant dies (Figs. 44–47).

7.3 Adaptations of the Main Organs

Whereas previous analyses and demonstrations have been carried out either in ecomorphological order (Chap. 5) or in phytosociological or ecological order (Chap. 6), the order of presentation in this chapter is morphological, starting with the shape of the entire plant, progressing through the stems, leaves, roots, and terminating in the diaspores.

7.3.1 The Overall Plant

The general appearance of the plant has, in sandy areas, a lot to do with its adaptation to the dune environment. As discussed in Section 2.2, plants function as obstacles to wind or air flow and hence influence sand mobility. Since the foremost component of sand mobility is the horizontal intensity of the wind, plant architecture is of great importance.

7.3.1.1 Canopy Density

Canopy density is an important adaptive property. It has a direct impact on wind velocity and hence on soil-particle mobility inside and around the plant. In their evaluation of the geomorphological bearing of plants on sand mobility, Thomas and Tsoar (1990) assume that a grass may create a windless zone inside the plant tuft. The canopy density modifies air flow within and near a plant's boundaries, and the relationship between vegetation cover and dune stability becomes complicated. Keep in mind that the movement of coarse sand is by creeping, and that of fine sand is mainly by saltation; therefore both are influenced by the energy of local winds. On the slipface of dunes all grain-size classes move by mass flow or avalanching (McKee 1982). Silt and clay are carried in suspension as aerosols (Tsoar and Pye 1987; Greely and Arvidson 1990) and, to a specific site by global and local wind storms. Deposition takes place when wind speed is minimal; it is often associated with rain or dew. Dense canopies function as traps for silt and clay (Danin and Yaalon 1981; Danin and Ganor 1991) by decreasing wind velocity. Some parts of plants also function as energy absorbers for airborne grains during the saltation process (Fig. 2), and thus stop movement. I suggest that three classes of canopy density can be recognized. The first includes very loose plants, such as *Cyperus sharonensis* (Fig. 68) and *C. capitatus*, which have only a very small impact on sand mobility. The grass in Fig. 82, *Elymus farctus*, had such a small influence

Fig. 82. A sparse *Elymus farctus* without a mound. The tips of leaves are so sparse that they have no impact on ripple direction

on wind velocity that it scarcely changed the direction of the ripples close to it. The same species under a different local situation has a profound impact on the direction of ripples (Fig. 4).

A second class of canopy density induces sedimentation inside the plant boundary. As a result of vigorous development of the plant tuft (Disraeli 1984), sand trapping becomes efficient and the plant builds a nebka. Plants of this type are *Stipagrostis scoparia*, *Ammophila arenaria*, *Ammophila breviligulata* (Disràeli 1984), *Calligonum comosum*, and *Scrophularia hypericifolia* (Fig. 83). Such canopies may trap fine sand and, under appropriate conditions, some silt and clay.

The third class decreases sand accretion inside the plant to minimal values, while an increased wind velocity at the plant's margins causes increasing wind erosion and the creation of a moat around the tuft (Fig. 8). Ash and

Fig. 83. *Scrophularia hypericifolia* with a mound accumulated as a result of plant impact on wind velocity. This plant has an intermediate canopy density

Wasson (1983) estimate the decrease of wind speed within the canopy to be 20% of the wind in the open area, and the increase near the plant, where moats develop, to be 120%. The floor of these depressions is rich in coarse sand, and even in particles larger than sand. Evidently, such plants trap fine sand, silt, and clay. Plants of this type include *Stipagrostis obtusa* and *S. ciliata.*

7.3.1.2 Plant Silhouette

Plant silhouette may have a considerable impact on wind flow and consequently on sand mobility. A plant which looks like an inverted cone directs wind flow from the canopy to its base and thereby causes an increase in wind velocity and erosion at ground level. Such a shape is found among trees and shrubs and mainly in those plants which do not produce shootborne roots, e.g., *Retama raetam* and *Haloxylon persicum.* Such a silhouette is rarely found in perennial grasses (e.g., *Stipagrostis drarii*).

Perennial grasses and shrubs which are adapted to grow in dunes have a loose global or cylindrical silhouette. This structure causes a decrease in wind velocity within the plant canopy, thus leading to the accretion of fine sand among the stems, and to the formation of nebkas. There is no increase in wind speed at the plant's margins, and thus there is no wind erosion in this microhabitat. Plants with this shape include species of *Calligonum*, and perennial grasses such as *Stipagrostis scoparia*, *S. pungens*, *S. acutiflora*, and *Ammophila arenaria.*

The combination of a cylindrical shape and dense canopy functions as an obstacle to wind flow and hence induces an increase in wind speed and consequent erosion near the plant. Wind velocity within the plant canopy is nearly zero, thus leading to the occasional trapping of sand, silt, and clay particles. Such plants are *Stipagrostis ciliata*, *S. obtusa*, and *Elymus farctus.*

Young plants of *Artemisia monosperma* have the shape of an inverted cone and thus do not interrupt sand mobility. Once they start building a nebka, plants of this species keep growing and may produce nebkas 2–3 m high above the dune surface.

Plants with extremely loose silhouettes, such as *Cyperus capitatus* in the coastal areas of Israel and Turkey, and *Cyperus sharonensis* (Fig. 68), barely function as an obstacle to sand mobility and allow sand movement to carry on as it would in an area devoid of plants.

7.3.2 Stems

As the main supporter of all other aboveground parts such as leaves, flowers, and fruits, the stem has to be well-adapted to sand accretion, sand deflation, and to injuries caused by airborne sand. Many plants have subterranean stems which also contribute to the ability to grow in desert dunes.

7.3.2.1 Aboveground Stems

The most obvious adaptation of plants to dunes with sand accretion is reactive growth by the production of shootborne roots. In addition, dormant buds above the roots sprout and give rise to new branches or tillers which eventually replace the buried stems. The stems of many desert psammophytes are viscid. Sand adheres to their surface and insulates them from the abrasive hazard of airborne sand. Leaves may have similar coatings (Sect. 7.3.3.2).

There are many stem-assimilants among shrubby desert psammophytes and among the grasses. The thick epidermis, thick cuticle and waxes, and the richness in sclerenchyma (Lyshede 1977, 1979; Fahn and Cutler 1992) are well known adaptations against drought. These may also protect photosynthesizing tissues from sand injuries. *Petalonyx thurberi*, a rare shrub of the Mojave Desert, California, has a thick corky bark on stems which are more than 5 mm in diameter. In a habitat with considerable sand movement (Sect. 6.8.2), the bark may serve to protect the stem from injuries and desiccation.

7.3.2.2 Subterranean Stems

Stems (geocorms) also serve as a bud bank. In some species, the subterranean bud bank can remain dormant but viable for several years. In most plants these buds have rigid outer scales which protect them from desiccation after being exposed to desert air. A bud bank permits reactive growth after deflation, the accretion of sand, and senescence of old stems and roots.

Many of the plants with geocorms have one or another mode of protection from desiccation: rigid stems; rich sclerenchyma with a thick epidermis; or are mostly covered by scales, leaf sheaths, or leaf bases. Some plants have a layer of ordinary hairs or colleters which assist in protection from abrasion when they are exposed. The hypocotyl, epicotyl, petioles, and the lower parts of the stems of seedlings and young plants of many desert psammophytes are protected from abrasion by affixed sand. Examples are provided by *Savignia parviflora* (Fig. 69), *Dicoria canescens* (Sect. 5.10.6) and its companion dicots in the Kelso and Owens Dunes - *Croton californicus*, *Cleome sparsifolia*, two species of *Gilia*, *Chamaesyce parryi*, *Abronia villosa*, and *Palafoxia arida.*

The first two types of geocorms listed below are plagiotropic, whereas the third is geotropic.

7.3.2.2.1 Rhizomes

Leafy rhizomes develop in plants with a dense canopy by intravaginal tillering. In grasses and other monocots, they resemble pachymorph rhizomes in bamboo and *Arundo donax.* In some plants, such as *Cyperus macrorrhizus* (Sect. 5.3.4), the plasticity of the leafy rhizome and its ability to elongate as a reaction to sand accretion sustain the plant in unpredictable desert sand dunes. Some desert dune plants, e.g. *Panicum turgidum* and *Pennisetum divisum*, have the ability to sustain slowly expanding growth by leafy rhizomes and to rapidly expanding growth by rhizodes. The latter may take place when soil moisture is above a certain threshold.

7.3.2.2.2 Rhizodes

Few species that are confined to desert dunes and require accumulating sand (Sect. 5.1) have plagiotropic rhizodes. *Panicum urvilleanum*, however, has a well-developed rhizode system with orthogeocorm tillers. The most successful nondesert pioneers, *Ammophila arenaria*, and *A. breviligulata*, can also expand by rhizodes and by orthogeocorms. Rhizodes may also serve as carbohydrate reserves, as in *A. breviligulata* (Seliskar 1994).

7.3.2.2.3 Orthogeocorms

All members of the ecomorphological group of species confined to areas of sand accretion have orthogeocorms. Orthogeocorms enable them to withstand sand accretion. The vertical cover of sand on the plant must be overcome by reactive vertical growth. *Stipagrostis* species and *Swallenia alexandrae* achieve such vertical growth by producing aerial tillers which turn into orthogeocorms and develop nodal roots when covered by sand. *Panicum urvilleanum* and *Ammophila arenaria* produce orthogeocormic tillers which carry leaves.

An important property of grasses and other monocots is the vertical elongation of orthogeocorm internodes by the activity of intercalary meristems. Internodes of *Ammophila arenaria* on orthogeocorms are longer, by two orders of magnitude, in sites of sand accretion than on stable sand (Sect. 5.9.1). Elongation of lower stem internodes also occurs in *Cutandia memphitica* (Sect. 5.8.5). The intercalary growth of internodes in grasses is well known (Fahn 1982) and seems to fit here as a mode for reactive growth.

7.3.3 Leaves

The leaves of desert dune plants do not differ much from those of nonpsammophytic desert plants. Various means of decreasing seasonal water loss by reducing the transpiration rate and a reduction of the leaf surface area are discussed in detail by Evenari et al. (1982). Leaves of desert dune plants typically display a thick cuticle. There is an extreme reduction of leaf area, almost up to complete absence, in stem-assimilants (Lyshede 1977, 1979; Fahn 1982; Fahn and Cutler 1992). Several of the stem-assimilants mentioned in Chapter 5 have been sectioned by Evenari et al. (1982) who display and discuss their typical anatomy. Many desert shrubs and semishrubs of desert areas have large leaves in winter and small ones in summer (Evenari et al. 1982; Orshan 1986b). Keshet et al. (1990) have examined several parameters of decreasing leaf area along the ecotone from maquis to steppe and desert in Israel. Many desert psammophytes follow the general trend of having smaller leaves than plants of more mesic conditions. Elongation of leaf sheaths in grasses is the most obvious reactive growth in leaves.

7.3.3.1 Leaf Shape

Elongation of the leaf sheath, which places the leaf blade above the sand, represents reactive growth of the leaf and an adaptation to sand accretion. This property has been found in *Panicum urvilleanum* (Sect. 5.1.6), and *Cutandia memphitica* (Sect. 5.8.5) and may occur in other grasses. Further elongation of mature leaves is achieved in grasses by intercalary meristems on leaf sheaths or at the base of the lamina. The cells of these meristems divide and contribute cells in both basipetal and acropetal directions (Fahn 1982; Mauseth 1988). The desert monocot *Carex pachystylis* was reported by Evenari et al. (1982) to be able to respond to the first winter rains by resuming the activity of 6-month-old dormant leaves. A few hours after the first rain, the cells at the leaf base start to grow and elongate, pushing their upper dry parts above ground. The newly formed tissue becomes green. The intercalary meristem at the base of the leaf lamina enables this fast growth to begin long before new leaves are formed at the apical meristem. Such rapid reactive growth seems to be an efficient adaptation to desert areas with an unpredictable availability of water. In a comparable way, the intercalary meristem of *P. urvilleanum* may enable it to overcome unpredictable sand accretion.

The laminas of many dune grasses and species of *Cyperus* are rolled up under conditions of water stress. By rolling up, the abaxial side of the leaf – protected by a thick shiny cuticle and without stomata – faces the harsh conditions of the desert air. The rolling may be accomplished by the contraction of bulliform cells (Fahn 1982) which occur only on the adaxial surface of the leaf and develop in lines parallel to the long axis of the leaf.

The leaves of many grasses have the ability to become desiccated from their tips in a basipetal direction, thus decreasing the transpiring area. Many grasses too, are capable of persisting throughout dry seasons, or even for years, with just green stems covered by green sheaths and dry laminas. Under extended extreme conditions, the leaf sheath may die too.

7.3.3.2 Leaf Indumentum

It is hard to define any one of the characters found in plants as an adaptation to a single environmental factor. The hairy leaves of coastal dune plants have been regarded as an adaptation to sea spray by Boyce (1954), and to high solar radiation by Fahn and Cutler (1992) and Barbour et al. (1993). Others consider hairs to be an adaptation to low humidity. According to this concept, humid air trapped among the hairs substantially decreases transpiration. I would recommend that experimental work be carried out to test my hypothesis that the hairiness of dune plants is an adaptation protecting them from the abrasive impact of airborne sand grains.

Many species growing in sandy areas have a mantle of sand on their aboveground parts; this I regard as a further measure of protection against sand abrasion. As discussed in Section 5.10.4, Fahn and Cutler (1992) consider this property to be an adaptation protecting the plant from solar radiation. This property is found in plants of desert as well as coastal and mesic dunes, and in many families. Several coastal-dune plants of moist areas, e.g., *Abronia maritima, Abronia umbellata, Calystegia soldanella*, and *Silene succulenta*, have sand on their stems even when they are growing below the soil surface; that is, they are preadapted to being exposed. As discussed in Section 5.2.2, *Artemisia monosperma* has hairy seedlings and glabrous adults with a sticky substance in the vicinity of the young buds.

7.3.4 Roots

Fahn (1982) lists the following properties of the root systems of plants as adaptations to desert conditions: (1) the form of the root system; (2) succulence of the roots; (3) development of thick bark; (4) sclerification of the cortical cells; (5) isolation of the vascular cylinder by periderm formation or by necrosis of the cortical parenchyma. However, since my main goal in this book is to identify special adaptations of desert dune plants, I review here the modes of reactive root growth posed by desert dune plants as well as special morphological features which may be associated with restraints imposed by a sandy habitat.

7.3.4.1 Shootborne Roots

The ability to produce adventitious roots as reactive growth to sand accretion seems to be one of the most important properties of plants confined to desert and nondesert dunes. In recent reviews of adventitious roots, Barlow (1986, 1994) and Groff and Kaplan (1988) recommended calling them "shootborne roots". The ability to produce roots from shoots covered by sand enables continued growth to take place at the soil surface. In other words, the plant moves vertically to the new soil surface. Vertical tillers covered with sand become orthogeocorms after they produce nodal roots. One of the best indicators for plants able to produce shootborne roots is the occurrence of large nebkas.

Among the grasses confined to desert sands are many species which do not produce nodal roots and have either poorly developed, or no axillary buds on their aerial tillers. They are normally found in stable sand sheets and rarely in sites of sand accretion. The ability to produce shootborne roots is not a common feature of all desert shrubby psammophytes. The most striking in their inability to do so are *Retama raetam*, *Hammada salicornica*, and *Anabasis articulata*. All three are plants after which plant communities of sandy areas have been named (Danin et al. 1964; Zohary 1973; Danin 1983).

A few trees are capable of producing shootborne roots. These include species of *Tamarix*, many of which reproduce by the vegetative propagation of accidentally detached branches. Such reproduction is common in desert wadis where a species of *Tamarix* may establish itself upstream from seeds and, later on, will propagate from natural cuttings downstream (Danin 1981b). *Prosopis glandulosa* in southern New Mexico (Campbell 1929) and in Death Valley, California, tolerates sand accretion in consequence of reactive root growth.

7.3.4.2 Rootborne Shoots

Several Old World Boraginaceae (Sect. 5.4) display a prominent type of reactive growth in sites with sand deflation. When deflation takes place, as a result of a local blowout or for other reasons, the succulent roots become exposed and rootborne shoots are produced. Even when the mother plant dies, its roots may for some time retain the ability to produce shoots.

7.3.4.3 Root System Morphology

The form of the root system is the first of the adaptations of roots to the desert that are listed by Fahn (1982). Shallow and wide ranging root systems of the plants of sandy areas are regarded by Zohary and Fahn (1952) and by Fahn (1982) as adaptations to a habitat where water is retained at the soil surface for only a short time. These investigators and others found that xylem vessels in vertical roots are narrower than those in horizontal roots. These vessels regulate suction pressure in the roots (Fahn 1982).

7.3.4.4 Special Functions of Roots

7.3.4.4.1 Absorption of Dew, Fog, and Condensed Water

Several authors have investigated the sources of water other than rainfall, that may be used by plants growing in extreme deserts. Vaadia and Waisel (1963) studied the direct absorption of dew by several desert plants growing in a sandy area south of Beer Sheva, and concluded that the amounts involved were negligible. They assumed that xerophytes, because of their thick cuticles, are less efficient than mesophytes in rapid foliar absorption of water during the night. *Trianthema hereroensis*, growing on sand dunes in the fog zone of Namibia, is one of the few desert plants shown to use fog efficiently (Louw and Seely 1982). Louw and Seely sprayed tritiated water on the top 1 cm of soil near *Stipagrostis sabulicola* plants and found efficient water absorption by the plant. It seems that dew can be absorbed by superficial root systems and may be an important source of water for some desert plants. Condensation of dew and fog on smooth leaves and the stems of perennial grasses may lead to water dripping directly on to the rhizosphere. Trunk flow is a well-known feature in the moisture and nutrient regime of plants in temperate areas and should be studied in desert dunes as well.

7.3.4.4.2 Rhizosheaths

The roots of many perennial grasses inhabiting desert sands are covered by persistent root hairs and sheaths of sand grains (Figs. 32, 34, 39–42) known as "rhizosheaths" (Volkens 1887; Price 1911; Troughton 1957; Wullstein et al. 1979; Wullstein and Pratt 1981; Wullstein 1991; Fahn and Cutler 1992). Volkens (1887) studied the anatomy and physiology of desert plants in Egypt, and described rhizosheaths as being an adaptation to desert dunes. He studied a few species of *Aristida* (now known as *Stipagrostis*), *Andropogon foveolatus* (= *Eremopogon foveolatus*), *Elionurus hirsutus* (= *Lasiurus scindicus*), and *Panicum turgidum*. The sand grains are attached to the root hairs by mucilage secreted from certain cells in the root tip. The whole outer layer of the root tip, up to a few millimeters away from the apex, looks like the palisade layer of a leaf and secretes the mucilage (Price 1911). Mucilage formation has been most extensively studied in the annual crop *Zea mays* (Vermeer and McCully 1982; McCully and Canny 1989). It was found, to be extracellular, histologically distinct from the mucilage at the epidermal surface, but similar to that produced by the root cap (Vermeer and McCully 1982). The mucilage is associated with detached cells, probably derived from the root cap, which remain alive in the soil sheath for a few days at least. The mucilage may be of a mixed origin, from free cells, root cap cells and bacteria. McCully and Canny (1989) do not recognize the rhizosheath of desert dune plants as a feature common to grasses in other environments. The developmental processes of the soil sheath on maize roots, as described by McCully and Canny (1989), involve deterioration of the sheath in the basal parts of the growing root and its development close to the apical part of the root. The situation in perennial desert dune grasses is that the rhizosheath persists for much longer. Volkens (1887) and Price (1911) believed that the rhizosheath promotes water absorption and protects the root from drought. They wrote that its protective function is analogous to that of the corky periderm in dicotyledonous stems. Wullstein et al. (1979) and Wullstein (1991) found the water content of the rhizosheaths to be four times that of control sand samples. Wullstein et al. (1979) found a relatively high rate of nitrogen fixation (tested by acetylene reduction) in the rhizosheaths of a few perennial grasses from sandy desert areas. They isolated heterotrophic nitrogen-fixing bacteria resembling *Bacillus polymorpha* from the roots of the following perennial grasses from sand dunes in the Great Basin of the United States: *Oryzopsis* (now *Achnatherum*) *hymenoides* (Sect. 5.6.3), *Agropyron dasystachyum* (now *Elymus lanceolatus*), *Stipa comata*, and *Aristida purpurea*. Wullstein (1991) continued his study of *Achnatherum hymenoides* (using the basionym *Stipa hymenoides*) and obtained similar results. He compared nitrogen fixation in plants derived from sand and having rhizosheaths, with plants derived from clayey soil and lacking rhizosheaths. There was no nitrogen fixation in plants without rhizosheaths. Nitrogen fixation near plants with rhizosheaths increased with an increasing availability of water. He concluded that during times when the dune was relatively dry, microbial activity would be maintained within the rhizosheaths where moisture conditions are more favorable. Moreover, nitrogen fixation may be specifically associated with rhizosheaths in spring or early summer. Rhizosheaths apparently enhance the uptake of several elements by *A. hymenoides* (Wullstein, in Harper and Pendleton 1993). One may look upon the rhizosheath as a site having special "nano-climatic" (smaller than "microclimatic") conditions which enhance the survival and activity of bacteria involved in nitrogen fixation. The idea of "nano-climatic" conditions was applied by Nienow et al. (1988) to the cryptoendolithic community of microorganisms in sandstones of the cold desert of Antarctica. The microorganisms in that extreme desert only survive in a very thin layer close to the rock surface where water is not frozen for a short time during the year and the temperature allows some activity. A similar situation is that of the nitrogen fixing bacteria which, in desert dunes, find appropriate living conditions only in rhizosheaths.

Persistent long root hairs were also noted in grasses by Metcalff (1960), who commented that attachment of sand grains to the roots is particularly obvious in dry regions. He did not relate the occurrence of rhizosheaths specifically to desert dunes, nor did he limit the definition of "dry" towards more mesic habitats. A study of the distribution of rhizosheaths has been carried out in Israel and Sinai in order to determine if there is a distinction between arid and mesic zones in this respect (Danin 1993, unpublished). Roots of perennial and annual grasses and of other monocots were examined from several sites within the climatic range of 20–500 mm mean annual rainfall. Plants were studied in the field and from herbarium material deposited in Jerusalem (HUJ). The results are presented in Table 4. Most sites were on sand sheets. A few specimens in M1 were collected from a mobile dune and one species was collected in M4 from a fresh-water swamp covered by sand. One can see from Table 4 that rhizosheaths are not restricted to the grasses of desert dunes. In Israel and Sinai they are found throughout the precipitation range and they are not restricted to mobile sand dunes. They occur in Gramineae, Cyperaceae, Juncaceae, and Liliaceae in sandy habitats. In California I have also found rhizosheaths on *Tiquilia plicata* (Boraginaceae; Sect. 5.3.7), *Sarcobatus vermiculatus* and *Atriplex polycarpa*, (Chenopodiaceae) and *Tamarix* cf. *ramosissima* in desert dune areas. *Calystegia soldanella*, which I examined in costal dunes near San Louis Obispo, California, had rhizosheaths on shootborne roots. Some annual grasses (Table 4 and Price 1911), *Carex pachystylis*, a few dicotyledons, and a few perennial grasses have long root hairs that entangle sand grains (TG in Table 4), but due to the absence or paucity of mucilage the sand grains fall off easily.

Seedlings of *Stipagrostis plumosa* (from S1), *S. scoparia* (from S3), and *Centropodia forskalii* (from M1) display different degrees of rhizosheath development between embryonic and nodal roots. The embryonic root is branched. Its root hairs are not dense and they entangle only a few sand grains. In contrast, nodal roots are not branched, are densely covered by root hairs, and have rather compact rhizosheaths.

Rhizosheaths also occur on the roots of the phreatophytes *Scirpus holoschoenus*, *Phragmites australis*, and *Juncus acutus*. These plants typically grow in sites with a high water table. Dense populations of these species are found in sandy habitats all along the Mediterranean coastal area of Israel, in depressions where the water table is as high as 1 m. There are sites and situations where the water table reaches the soil surface, making a temporary swamp. The roots of these plants pass through dry and wet layers of sand. Where they pass through dry sand their roots have rhizosheaths.

To conclude, rhizosheaths occur on perennial grasses and other monocots (but rarely on dicots) growing in sandy soils. They are found in many genera

Table 4. Distribution of rhizosheaths in Israel and Sinai

Perennial grasses	A1	A2	M1	M2	M3	M4	N1	N2	S1	S2	S3	S4	S5
Ammophila arenaria			TG		TG								
Centopodia forsskalii		RS	RS	RS	RS								
Elymus farctus						RS							
Hyparrhenia hirta			TG										
Imperata cylindrica						TG							
Lariurus scindicus	RS	RS								RS		RS	
Panicum turgidum	RS	RS	RS										
Pennisetum ciliare	RS												
P. divisum			RS			RS		RS				RS	
P. elatum													RS
Phragmites australis				RS									
Stipa pellita						RS							
Stipagrostis acutiflora										RS			
S. ciliata							RS			RS			RS
S. hirtigluma									RS				
S. lanata		RS	RS										
S. obtusa						RS							
S. plumosa	RS					RS	RS						
S. raddiana		RS											RS
S. scoparia			RS				RS			RS			
Annual grasses													
Avena wiestii						TG							
Bromus tectorum						TG							
Corynephorus divaricatus			TG		TG								
Ctenopsis pectinella			TG										
Cutandia memphitica			TG		TG		TG						
Rostraria smyrnacea			TG										
Stipa capensis	TG					TG	TG						
Trisetaria linearis			TG		TG								
Other families													
Asphodelus ramosus							RS						
Carex pachystylis						TG							
Cyperus capitatus				TG	RS								
C. conglomeratus										RS			
C. macrorrhizus			RS	RS					RS				
Juncus acutus					RS	RS							
J. bufonius					TG								
Launaea resedifolia		TG											
Polygonum palaestinum		TG											
Scirpus holoschenus					RS	RS							
Senecio joppensis		TG											

Mean annual precipitation in the sites studied: A1, A2 = 30–50 mm; M1, M2, M3, M4 = 450–500 mm; N1, N2 = 90–100 mm; S1, S2, S3 = 20–50 mm; S4, S5 = 70–100 mm. Root type: RS = rhizosheaths present; TG = sand attached by long tangled root hairs; A1 = Arava, near Hazeva; A2 = near Elat; M1 = Mediterranean coast 10 km S of Ashkelon; M2 = Nizzanim; M3 = Netanya; M4 = 7 km E of Haifa; N1 = Negev, 10 km SE of Dimona; N2 = 27 km SSE of Beer Sheva; S1 = Sinai, Wadi Mukkatab; S2 = Sinai, Dahab to Sharm el Sheikh; S3 = Sinai, 15 km NW of Mitla Pass; S4 = Sinai, Gebel Maghara; S5 = Sinai, Gebel Kheil near Sarabit el Khadem but are absent in others. There is no noncontroversial proof that rhizosheaths are an adaptation to nutritional stress (as suggested by Wullstein et al. 1979) or to dry conditions in sandy habitats, nor that they have anything to do with nutrition or drought stress in sands. However, one has to bear in mind that perennial grasses are the leading plants of mobile dunes in desert areas with more than 70 mm mean annual rainfall. It appears that rhizosheaths are important for plants in many sandy areas which are relatively dry and poor in nutrients, and they are an essential part of the adaptive syndrome of the plants which require sand accumulation (Sect. 5.1). Being part of the morphological syndrome of the desert-dune colonizers, rhizosheaths and their N_2-fixing bacteria should be regarded as a kind of "mutualism", in the way in which it is defined, described, and discussed by Barbour et al. (1987).

7.3.4.4.3 Mycorrhizae

The roots of most plant species are associated with one or more species of mycorrhiza-forming fungi. Nonmycorrhizal plants are unusual (Jackson and Mason 1984). The fungus and its host support one another in mutualistic symbiosis. In his review, Mosse (1973) concludes that vesicular-arbuscular mycorrhizae improve plant nutrition and vigor, especially in nutrient-deficient soils.

Mycorrhizae have been studied in coastal dunes in several countries. A few studies on the grasses of coastal dunes demonstrate mycorrhizal association with *Ammophila breviligulata* and all its dicot companions in Rhode Island (Koske and Halvorson 1981), *Ammophila littoralis* in Italy (Puppi et al. 1985; Pacioni et al. 1985), with *Uniola paniculata* in Florida (Sylvia 1985), and with other grasses in India (Sabharwal and Mukerji 1985). Spores of endomycorrhizal fungi were found in coastal sand dunes of New South Wales, Australia (Koske 1975), their density being higher in stable than in mobile sand. Mycorrhizal fungi may contribute to stabilization by linking sand grains in aggregates with fungal hyphae (Sutton and Sheppard 1976).

Studies of the resistance of *Ammophila breviligulata* to drought (Koske and Halvorson 1981; Smith 1980; West 1994) have shown that mycorrhizae improve plant establishment and may serve a similar role in desert dune plants.

Stable dunes in the western Negev, with relatively high quantities of silt and clay, support several species of *Helianthemum* which have ectomycorrhizal relationships with tuberous fungi, commonly known as truffles (Binyamini 1980; Roth-Bejerano et al. 1990; De Malach et al. 1995). The identity of the mycobiont, and its ecological value to the vascular plant are not yet known. On sandy soils of the northern and western Negev the vascular plants are *H. sessiliflorum*, *H. kahiricum*, *H. stipulatum*, *H. ledifolium*, and in the southern Negev it is *H. lippii*.

7.3.5 Seed Dispersal

In this chapter I shall list the specialized modes of seed dispersal of the prominent desert dune plants.

7.3.5.1 Dragged Diaspores

The most specialized growth forms in the desert dunes ecosystem are those of the species which require sand accretion (Sect. 5.1). The diaspores of these species must arrive at sites where sand accretion takes place. The dispersal unit of *Stipagrostis scoparia* is a relatively heavy caryopsis subtended by a glabrous palea and lemma which terminate in three nearly equal plumose awns. These awns are straight when immature and curve backwards when mature and ready for dispersal (Fig. 23a). The dispersal unit therefore has a parachute-like architecture and looks like a common pappose diaspore of the Asteraceae (Compositae; Sheldon and Burrows 1973). During a strong wind I have seen such diaspores dragged by the wind at a high speed on the dune surface and not gliding in the air. This kind of movement does not fall into the classes of movement discussed under "plumed diaspores" by Pijl (1972), or "guide parachute" and "drag parachute" by Hoerner (1965), Murray (1986), and Niklas (1992). The diaspores of *S. scoparia* are dragged on the bare dunes like a paratrooper who has landed but not disconnected himself from his parachute. Murray's (1986, p 37) illustration and equations concerning diaspores with parachutes, show clearly that if seed weight is high, the aerodynamic system leads to dragging on the ground. However, exceptionally strong vertical convection may cause gliding. Dragged diaspores come to a rest on the leeward slopes of dunes where wind velocity approaches zero. Diaspores landing there may be covered by sand grains arriving by saltation, creeping, or sliding. Diaspores are spread all along the leeward slope. In this way the diaspores of *S. scoparia* arrive at appropriate safe sites and remain in the dune. As the leeward slope slowly moves forward, new diaspores are deposited in the new leeward slope. Consequently, the entire dune area becomes a seed bank at all depths (Sect. 7.3.5.6). Seeds remain in the system of mobile sand which is the sole habitat of this species, protected from granivorous animals. The safe leeward sites constantly change their position in space but always remain available for safe diaspore deposition for *S. scoparia* and other species whose seeds are similarly dragged by the wind. Even if the sand continues to move and seeds which were close to the dune surface become exposed, there are additional seeds in the deeper sand layers. Exposed diaspores may move with strong winds by a mode of creeping, as large grains of sand do (Sect. 2.2.1). Gliding diaspores of other species may be deposited on the leeward sites and be protected there as well. On the leeward slopes of mobile dunes in Namibia, I saw many diaspores of *Stipagrostis ciliata*, even though there were hardly any living adult plants of this species in the dune system that particular year.

It should be noted that the diaspore tip of *S. scoparia* is pointed but not hairy, and the base of the awn is not twisted. I believe that the absence of these hairs must be a derivative adaptation since most *Stipagrostis* spp. and many other genera of grasses have hairs or minute teeth on the diaspore surface directing it into the ground. Such hairs or teeth would be an obstacle to the diaspore dragging on the sand. The diaspore of this species is "programmed" to penetrate the soil in a different way than that of other grass species.

The other common type of diaspore in the genus *Stipagrostis* has a plumose central awn with long hairs and two lateral awns which are glabrous or with short hairs (Fig. 23b). The caryopsis is smaller and lighter than the previous type. Such a diaspore glides in the air and is deposited on the ground when wind ceases. Many have a palea and lemma with hairy bases; these hairs are stiff and turned upwards guiding the diaspores into cracks in the soil and holding them there (Pijl 1972). The awn's base is often twisted and functions as a mechanism which drills the seed into the ground by twisting and untwisting with changes in air humidity (Stebbins 1971).

The dispersal syndrome of most *Stipagrostis* species which require sand accretion in desert dunes is, therefore, seed-setting throughout the year; awns which constitute a drag-parachute; relatively heavy caryopsis; and an absence of stiff hairs at the tip of the diaspore. *Stipagrostis scoparia*, *S. sabulicola*, *S. pennata*, *S. karelinii*, and *S. lutescens* all have these properties. Species of this ecomorphological type (Sect. 5.1) with plumed diaspores include *S. amabilis S. acutiflora*, and *S. pungens.* Of nearly 60 species of *Stipagrostis* that I studied from herbarium sheets every one which had drag-parachute awns was confined exclusively to mobile sand dunes (Danin 1994).

7.3.5.2 Rolling Diaspores

Another mode of directed seed dispersal is that of *Calligonum* spp. Most of the species of this genus are desert psammophytes but they are tolerant to a wide range of mobility. For example, *C. comosum* occurs in mobile dunes, in stable sand, and in sand under deflation. Its diaspore is a woody achene with 10–16 rows of bristles. A heavy diaspore with short bristles cannot be classified as a gliding diaspore. The bristles are too soft and straight to function in ectozoochory. The diaspores are rolled on the dune surface until they arrive at the leeward side of a dune. There, as described in the case of *S. scoparia*, they are buried in the dune at various depths. According to Boissier (1879) and Komarov (1936) the genus contains dozens of species with setose fruits (Fig. 84a) and dozens with winged fruits (Fig. 84b). Both types of fruits roll on the sand surface.

Rolling is also the mode of dispersal of many rare perennial grasses which require sand accumulation and are confined to desert dunes in southwestern North America. The most prominent of these is *Swallenia alexandrae*, an endemic perennial grass of Eureka Dunes, California. The diaspore is either

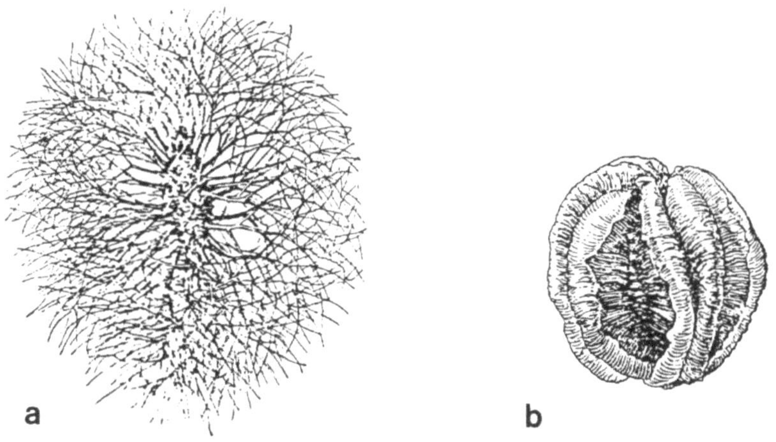

Fig. 84a,b. Fruits of *Calligonum* species adapted to rolling on the ground: **a** *C. eriopodum*; **b** *C. lanciculatum*. (Komarov 1936)

the naked caryopsis or it contains a caryopsis, palea, and lemma covered by long hairs. It may be disseminated near the mother plant or rolled on the sand surface as described in the case of *Calligonum. Panicum urvilleanum* is another dune plant (Sect. 5.17) with a rounded diaspore. The glumes, palea, and lemma subtending the caryopsis have long hairs which assist in rolling the heavy diaspore on the dune surface. *Achnatherum hymenoides* is a common dune grass of the Great Basin (Sect. 5.6.3) with a rounded diaspore; the lemma is densely covered with long hairs but the palea is glabrous. *Hilaria rigida*, an uncommon dune plant of the Great Basin (Sect. 5.2.5), has a hairy rounded spiklet-diaspore which appears to be adapted for rolling too.

The dispersal unit of *Haloxylon persicum* and of many other shrubby chenopods in the genera *Salsola*, *Hammada*, and *Anabasis*, is a one-seeded fruit subtended by five sepals each armed with a lateral wing. Seed dispersal is by wind. I think that dispersal is by rolling on the dune surface, rather than by gliding. Diaspores arriving on the leeward side of the dune are covered by sand and become incorporated into the seed bank of the dune, as described earlier.

7.3.5.3 Creeping or Bouncing Diaspores

Seeds of *Plantago cylindrica* and *Eremobium aegyptiacum* – annual psammophytes of the extreme desert of Israel and Sinai – seem to be adapted for dispersal in dunes in a simple way. The diaspore in each case is a small yellow to yellow-brown seed with an epidermal layer of cells which become mucilaginous when wetted with water. The seeds are larger than sand grains, but may weigh about as much as large grains. They move together with sand grains by saltation, creeping, or sliding (Fig. 2).

Diaspores of *Ifloga spicata*, an annual psammophyte of the desert and coastal dunes of the eastern Mediterranean and Saharan desert, are small. Those situated on the receptacle have a rudimentary pappus composed of 2–6 bristles which can hardly serve as a glide-assisting apparatus. The diaspores at the axil of the inflorescence bracts are devoid of a pappus. I think both kinds are probably dispersed by bouncing, in a similar way to those of *Plantago* and *Eremobium*. These seeds are mixed into the sand at various depths. They are protected from granivors by their similarity in colour and size to sand grains, and by burial.

7.3.5.4 Conclusions and Comparison with Seed Dispersal Among Island Plants

Three specific modes of seed dispersal are found in dune plants: dragging, rolling, and bouncing. The directed wind dispersal of desert dune species enables them to remain in their most appropriate sites and avoid being carried off the dune to unsafe sites. The dispersal syndromes of desert dunes plants resemble those of island plants as discussed by Carlquist (1974), who concluded that island groups of plants have lost dispersibility. For example, dune species of the genus *Stipagrostis* which belong to the *S. scoparia* type are dispersed by wind and most of them have a heavy, parachute-like diaspore. This nongliding diaspore, as compared with the gliding diaspores of nondune species of *Stipagrostis*, represents a loss of dispersibility. However, since these obligate dune plants have no biological advantage except in sites with sand accretion, long-distance dispersal may mean a loss of reproductive material. They must remain in their mobile dune island habitat. Thus, their seed dispersal syndrome is directed towards remaining in the dunes. *Swallenia alexandrae* is another example for the desert dune–island link. Its diaspore is either a naked caryopsis, without any assisting apparatus, or a caryopsis with hairy palea and lemma constituting a rolling diaspore. Large areas of nonsandy habitat isolate Eureka Dunes from all other dunes. No reproductive efforts are wasted outside the small "sand island" to which this endemic species is restricted.

8 Comparison Between Plants of Desert Dunes, Extreme Desert, and Coastal Dunes

8.1 Comparison of Desert and Coastal Dunes

The principal differences between the plants of desert and coastal dune areas are that in many coastal dune systems, plants have to withstand the stresses of salt-spray and inundation by sea water in addition to that of sand mobility (Oosting and Billings 1942; Boyce 1954; Barbour et al. 1973, 1985; Chapman 1976). Coastal dunes are populated by plant species different from those of the desert. In contrast to the large proportion of grass species in desert dunes, coastal dunes are floristically more diverse. For example, the coastal dunes of East London, South Africa, are rich in species of Asteraceae and Goodeniaceae, and only a small proportion of species are Poaceae (Burns and Lubke 1986). Similarly, sandy coasts of California are populated with species of Asteraceae, Euphorbiaceae, Nyctaginaceae, and Onagraceae (Barbour et al. 1973, 1985).

Chapman (1976) indicates that the plant succession on coastal dunes of Europe and North America is mainly of hemicryptophytes, and the climax is of phanerophytes. In northern Israel the succession begins with perennial grasses, continues with chamaephytes, and the climax is of phanerophytes (Kutiel et al. 1979/80; Danin and Yaalon 1982). In contrast, few phanerophytes are present in desert sand dunes. Mobile sands in deserts are dominated by grasses, whereas stable sands are populated by chamaephytes.

In moister coastal areas the rates of plant-dependent processes are faster and the accumulation of organic matter as humus during plant succession is prominent. *Ammophila arenaria* (Sect. 5.9.1) is the principal species adapted to accumulating sand in the coastal dunes of Israel, northeast Sinai (Danin 1983; Danin and Yaalon 1982), most of the Mediterranean countries, the British Isles, and western Europe (Ranwell 1972; Corre 1975; Chapman 1976; Huiskes 1979; Gehu 1985; Doing 1986), and California (where it was introduced). It has adaptations for accumulating sand which are similar to those of *Stipagrostis* species (Sect. 5.1; i.e., it has perennial but short-lived stems and roots which develop from buried nodes, and it has rhizodes that grow into leeward nebka). Moreover, it shares many growth features with *Panicum urvilleanum* of the Mojave Desert. The floristic assortment which has developed throughout the evolutionary process has been diverse; but the common solution, and thus the convergent property, which enabled grasses to function as pioneer colonizers of the dunes in desert and coastal areas is the ability to produce orthogeocorms.

The stability of coastal dunes in mesic areas is achieved by a complete cover of vegetation. In the Eastern Cape, South Africa, i.e., near Port Alfred and Alexandria, tall dunes (termed "precipitation dunes" Cooper 1958) develop at the contact zone between mobile dunes and coastal shrubby vegetation ("coastal thickets" sensu Lubke et al. 1988). There are a few woody species, such as *Brachylaena discolor*, in the thickets, which are able to grow above the accumulating sand, thus forming large nebkas which have been termed "ascending accretion dunes" (Tinley 1985). The shrubs or trees that build such dunes grow to a height of several dozen meters, thus preventing further movement of the dune system inland. In coastal areas of South Africa which have a high rate of sand accretion and high energy winds, the grasses growing on mobile dunes play only a minor role in stabilizing the sand. I assume that the tall phytogenic dunes are responsible for most stabilization. The mesic climate, deep roots anchored in mature soils, and probably shootborne roots, enable trees and shrubs (as in the Mediterranean coasts of Turkey and Israel - Sect. 5.9.10) to create these precipitation dunes parallel to the coast and hence influence the entire dune system.

Cyanobacteria develop in coastal dunes (Ancker et al. 1985; Nokrian, 1988; de Winder et al. 1989; Danin et al. 1989) but their role in the final stages of succession is not so crucial as it is in desert ecosystems. Mosses also contribute to the stabilization of coastal dunes as first noted in Europe (Chapman 1976) and in Australia (Moore and Scott 1979), and later observed in California (Danin et al. 1995). In coastal Australia moss species were found to be able to grow through an accumulation of 3–4 cm/year of sand (Moore and Scott 1979). The distribution of mosses in the coastal dunes of Israel and California is associated with the presence of humus, whereas in desert dunes mosses do not seem to be influenced much by the distribution of organic matter. The ecological role of mosses in desert and coastal dunes deserves further investigation.

The proximity of the dunes to the sea brings minerals to the nutrient-poor ecosystem. There is also a constant supply of organic fragments of dead marine organisms, algae, plankton, and the excretion of sea and coastal birds. Such resources do not exist in desert dunes.

The quantity of rainfall is of great significance in influencing not only the moisture regime but also soil salinity (Danin 1981a). In European slacks (interdune depressions) aquatic plants such as *Potamogeton* and *Zanninchellia* occur where fresh water is available all the year round. Other slacks, less wet, have *Juncus*, *Salix*, *Carex*, and *Scirpus* species (Chapman 1976). In the coastal zone of Israel (such as in the Akko–Haifa area), there are interdune salt marshes dominated by species of *Arthrocnemum*, *Sarcocornia*, *Limonium*, and *Tamarix*. A similar situation has been described by Corre (1975) from the Camargue of southern France. Most of the sandy areas along the coast of Israel are not influenced by a high water table. They are covered by seral communities influenced by the type of substrate beneath the sand and by the climatic conditions (Danin and Yaalon 1982). There is a park-like woodland of the *Ceratonia siliqua–Pistacia lentiscus* community and related seral communities in the northern part of the country, and shrubland in the south.

In northern Sinai, northeast of El 'Arish, there is a high water table of fresh water which is used for date palm cultivation (Danin 1983). In much of the terrain between El 'Arish and Port Said there are large areas where a salty water table lies close to the surface. The interdune salt marshes are dominated by the hydrohalophytes *Halocnemum strobilaceum*, *Arthrocnemum macrostachyum* (= *A. glaucum*), *Sarcocornia* spp., and *Suaeda* spp., whereas depressions where water is less salty are dominated by *Juncus arabicus* and *Phragmites australis* (Danin 1983). In drier areas, such as along the Gulf of Suez, the sand dunes are devoid of vegetation whereas, as a result of an underground flow of fresh or saline water, the interdune areas are covered with halophytic coastal vegetation (Danin 1981a, 1983; Evenari et al. 1985). The dominants are *Halocnemum strobilaceum*, *Arthrocnemum macrostachyum*, *Nitraria retusa*, *Limonium axillare*, and *Suaeda vermiculata.*

8.2 Morphological Differences Between Dune Plants of Desert and Coastal Areas

Are there any substantial morphological or ecomorphological differences between the plants of desert dunes and those of coastal dunes? The physiological differences related to the different input of nutrients and sea spray are not evident in the morphology of plants. The pioneer grasses of desert (Sect. 5.1) and coastal areas (Sects. 5.9.1–5.9.4) share ecomorphological traits, but none has a bimodal distribution. None of the traits which in previous chapters have been regarded as adaptations to dunes are endemic to desert areas. Moreover, in Israel and Sinai, a few plant species which are common to both areas, and differ at the ecotypic level, include members of several ecomorphological groups. Similar bimodal species, which are discussed in Section 5.2, are *Artemisia monosperma*, *Panicum turgidum*, and *Pennisetum divisum*. A representative of Section 5.3 is *Centropodia forskalii*, and close vicariants of this group are *Stipagrostis plumosa* in the desert and *S. lanata* in the coastal dunes. Representatives of Section 5.4 are *Moltkiopsis ciliata* and *Echiochilon fruticosum*, of Section 5.5 *Retama raetam*, and of Section 5.8, *Ifloga spicata* (recorded in Sect. 5.10.1) and *Lotus halophilous.* In California, *Croton californicus*, which belongs to the ecomorphological group of Section 5.5, grows in both coastal and desert dunes.

All perennial grasses inhabiting desert dunes have rhizosheaths, whereas most grasses of coastal dunes do not posses this property. However, rhizosheaths do exist on the shootborne roots of a few dicots in coastal dunes near San Louis Obispo, California. More research on the function and distribution of rhizosheaths is required.

8.3 Morphological Evidence for Delimiting Dunes of Extreme Desert and Desert

The boundary between "desert" areas and "extreme desert" areas is difficult to determine. However, the presence or absence of pioneer species which tolerate and require sand accumulation (Sect. 5.1) should be considered. These are perennial grasses, and in the "Old World" many of them belong to the genus *Stipagrostis*. The fact that they grow in a certain site indicates their requirements. This means that the amount of water accumulated in the sand is sufficient for their perennial growth. In areas of lower rainfall, where they do not grow, only annual species occupy the dune slopes and then only in wet years. Whereas in habitats other than dunes the pattern of spatial distribution changes (Monod 1931) from "diffuse" in wet to "contracted" in dry zones, the plants of Section 5.1 are direct indicators. In Israel and Sinai the representative of Section 5.1 is *Stipagrostis scoparia*. It grows in areas of diffuse vegetation (Danin 1983; D_1 in Fig. 14) and is absent from drier areas. Similarly, in California, Eureka Dunes and the Kelso Dunes support perennial grasses (Sects. 6.7.4, 6.8.3), whereas the extreme-desert dunes of Death Valley are populated by either annuals in wet years or by phreatophytes which use below-sand water.

Another way to look at one of the ecomorphological traits is to evaluate its indicative value. A diaspore such as that illustrated in Fig. 23a, found in situ, indicates (1) it is derived from the "Old World", (2) it is derived from mobile sand dunes, and (3) the amount of mean annual rainfall in that site is 70–250 mm.

Appendix
List of Higher Plant Names with Authors, Synonyms, and Family Name

Plant names with = sign before the name are synonyms or names that were commonly used for other plants in the past even if not strict taxonomic synonyms.

Abronia maritima S. Watson [Nyctaginaceae]
Abronia umbellata Lam. [Nyctaginaceae]
Abronia villosa S. Watson [Nyctaginaceae]
Acacia albida Delile [Mimosaceae]
Achillea millefolium L. [Asteraceae]
Achnatherum hymenoides (Roemer et Schultes) Barkworth [Poaceae]
Adonis dentata Delile [Ranunculaceae]
Aeluropus lagopoides (L.) Twaites [Poaceae]
Agathophora alopecuroides (Delile) Bunge [Chenopodiaceae]
Agropyron dasystachyum (Hooker) Scrib. [Poaceae]
= *Agropyron junceum* (L.) Beauv. (see *Elymus farctus*)
Ambrosia chamessonis (Less.) E. Greene [Asteraceae]
Ambrosia dumosa (Gray) Payne [Asteraceae]
Ammochloa palaestina Boiss. [Poaceae]
Ammophila arenaria (L.) Link [Poaceae]
Ammophila breviligulata Fern. [Poaceae]
Anabasis articulata (Forssk.) Moq. [Chenopodiaceae]
Anastatica hierochuntica L. [Cruciferae]
Anchusa negevensis Danin [Boraginaceae]
Andropogon foveolatus Delile [Poaceae]
Anthemis melampodina Delile [Asteraceae]
= *Aristida acutiflora* (see *Stipagrostis*)
= *Aristida amabilis* (see *Stipagrostis*)
= *Aristida ciliata* (see *Stipagrostis*)
= *Aristida drarii* (see *Stipagrostis*)
= *Aristida hirtigluma* (see *Stipagrostis*)
= *Aristida karelinii* (see *Stipagrostis*)
= *Aristida lanata* (see *Stipagrostis*)
= *Aristida lutescens* (see *Stipagrostis*)
= *Aristida obtusa* (see *Stipagrostis*)
= *Aristida pennata* (see *Stipagrostis*)
= *Aristida plumosa* (see *Stipagrostis*)
= *Aristida pungens* (see *Stipagrostis*)
Aristida purpurea Nutt. [Poaceae]
= *Aristida sabulicola* (see *Stipagrostis*)
= *Aristida scoparia* (see *Stipagrostis*)
Artemisia monosperma Delile [Asteraceae]
Artemisia tridentata Nutt. [Asteraceae]
= *Arthrocnemum glaucum* (see *A. macrostachyum*)
Arthrocnemum macrostachyum (Moric.) Moris Delponte [Chenopodiaceae]
Arundo donax L. [Poaceae]
= *Asphodelus microcarpus* Salzm. et Viv. (see *A. ramosus*)
Asphodelus ramosus L. [Liliaceae]
= *Asthenatherum forskalii* (Vahl) Nevski (See *Centropodia*)
Astragalus annularis Forssk. [Fabaceae]
Astragalus camelorum Barbey [Fabaceae]
Astragalus caprinus L. [Fabaceae]
Atriplex polycarpa (Torrey) S. Watson [Chenopodiaceae]
Avena wiestii Steudel [Poaceae]

Baccharis pilularis DC. [Asteraceae]
Bassia muricata (L.) Ascherson [Chenopodiaceae]
Brachylaena discolor DC. [Asteraceae]
Bromus fasciculatus C. Presl [Poaceae]
Bromus tectorum L. [Poaceae]

Cakile maritima Scop. [Cruciferae]
Calligonum comosum L'Her. [Polygonaceae]
Calligonum eriopodum Bunge [Polygonaceae]
Calligonum lanciculatum N. Pavl. [Polygonaceae]
Carduus getulus Pomel [Poaceae]
Calystegia soldanella (L.) R. Br. [Convolvulaceae]
Carex pachystylis J. Gay [Cyperaceae]
Centaurea aegialophila Wagenitz [Asteraceae]
Centropodia forskalii (Vahl) Cope [Poaceae]
Ceratonia siliqua L. [Caesalpinaceae]

Chamaesyce parryi (Engelm.) Rydb. [Euphorbiaceae]
Chilopsis linearis (Cav.) Sweet [Bignoniaceae]
Chondrila juncea L. [Asteraceae]
Chrysothamnus nauseosus (Pall.) Britton. [Asteraceae]
Cleome sparsifolia S. Watson [Capparaceae]
Colchicum ritchii R. Br. [Liliaceae]
= *Coldenia nuttalii* Hooker (see *Tiquilia*)
= *Coldenia plicata* (Torrey) Cov. (see *Tiquilia*)
Convolvulus lanatus Vahl [Convolvulaceae]
= *Convolvulus soldanella* L. (see *Calystegia)*
Coridothymus capitatus (L.) Reichenb. fil. [Lamiaceae]
Cornulaca monacantha Delile [Chenopodiaceae]
Corynephorus canescens Beauv. [Poaceae]
Corynephorus divaricatus (Pourret) Breistr. [Poaceae]
Croton californicus Muell. [Euphorbiaceae]
Crucianella membranacea Boiss. [Poaceae]
Ctenopsis pectinella (Delile) De Not. [Poaceae]
Cutandia memphitica (Sprengel) K. Richter [Poaceae]
Cyperus capitatus Vandelli [Cyperaceae]
Cyperus conglomeratus Rottb. [Cyperaceae]
Cyperus macrorrhizus Rottb. [Cyperaceae]
= *Cyperus mucronatus* (L.) Mabille (see *C. capitatus)*
Cyperus sharonensis Danin et Kukkonen [Cyperaceae]

= *Dalea polyadenia* Watts (see *Psorothamnus)*
= *Danthonia forskalii* (Vahl) R. Br. (see *Centropodia)*
Dicoria canescens A. Gray [Asteraceae]
Distichlis spicata (L.) E. Greene [Poaceae]

Echinops philistaeus Feinbrun et Zohary [Asteraceae]
Echiochilon fruticosum Desf. [Boraginaceae]
= *Elionurus hirsutus* (Vahl) Bentham (see *Lasiurus scindicus)*
Elymus farctus (Viv.) Melderis [Poaceae]
Elymus lanceolatus (Scribn. & Sm.) Gould [Poaceae]
Ephedra alata Decaisne [Ephedraceae]
= *Ephedra alte* C.A. Meyer (see *E. aphylla)*
Ephedra aphylla Forssk. [Ephedraceae]
Ephedra strobilacea Bunge [Ephedraceae]
Eremobium aegyptiacum (Sprengel) Boiss. [Cruciferae]
Eremopogon foveolatus (Delile) Stapf [Poaceae]
Eriogonum latifolium Smith [Polygonaceae]
Erodium crassifolium L'Her. [Geraniaceae]
Erodium laciniatum (Cav.) Willd. [Geraniaceae]
Erucaria pinnata (Viv.) Taeckh. et Boulos [Cruciferae]
= *Erucaria uncata* (Boiss.) Ascherson et Scweinf. (see *E. pinnata*)
= *Euphorbia parryi* (see Chamaesyce)

Fagonia arabica L. [Zygophyllaceae]
Fagonia bruguieri DC. [Zygophyllaceae]
Fagonia glutinosa Delile [Zygophyllaceae]
Fagonia mollis Delile [Zygophyllaceae]
Fagonia sinaica Boiss. [Zygophyllaceae]
Filago desertorum Pomel [Asteraceae]

Gagea dayana Chodat et Beauverd [Liliaceae]
Gilia sp. [Polemonaceae]
Gymnocarpos decander Forssk. [Caryophyllaceae]
Gypsophylla viscosa Murray

Halocnemum strobilaceum (Pallas) MB. [Chenopodiaceae]
Hammada salicornica (Moq.) Iljin [Chenopodiaceae]
Haloxylon ammodendron Bunge [Chenopodiaceae]
Haloxylon persicum Bunge [Chenopodiaceae]
= *Haloxylon salicornicum* (Moq.) Boiss. (see *Hammada*)
Haplophyllum tuberculatum (Forssk.) Ad. Juss. [Rutaceae]
Helianthemum kahiricum Delile [Cistaceae]
Helianthemum ledifolium (L.) Miller [Cistaceae]
Helianthemum lippii (L.) Dum.-Courset. [Cistaceae]
Helianthemum sessiliflorum (Desf.) Pers. [Cistaceae]
Helianthemum stipulatum (Forssk.) C. Chr. [Cistaceae]
Heliotropium arguzoides Kar. et Kir. [Boraginaceae]
Heliotropium digynum (Forssk.) C. Chr. [Boraginaceae]
= *Heliotropium luteum* Poiret (see *H. digynum*)

Hilaria rigida (Thunb.) Bentham [Poaceae]
Hippocrepis areolata Desv. [Papilionaceae]
Hyparrhenia hirta (L.) Stapf [Poaceae]

Ifloga spicata (Forssk.) Schultz Bip. [Asteraceae]
Imperata cylindrica (L.) Raeuschel [Poaceae]

Juncus acutus L. [Juncaceae]
Juncus arabicus (Ascherson and Buchenau) Adamson [Juncaceae]
Juncus bufonius L. [Juncaceae]

Larrea tridentata (DC.) Coville. [Zygophyllaceae]
= *Lasiurus hirsutus* (Vahl) Boiss. (see *L. scindicus*)
Lasiurus scindicus Henrard [Poaceae]
Launaea resedifolia (L.) O. Kuntze [Asteraceae]
Launaea tenuiloba (Boiss.) O. Kuntze [Asteraceae]
Limonium axillare (Forssk.) Kuntze [Plumbaginaceae]
Linaria haelava (Forssk.) Delile [Scrophulariaceae]
Linaria tenuis (Viv.) Sprengel [Scrophulariaceae]
= *Lithospermum callosum* Vahl (see *Moltkiopsis*)
Lobularia arabica (Boiss.) Muschler [Cruciferae]
Lotus halophilous Boiss. et Spruner [Papilionaceae]
Lycium schweinfurthii Dammer [Solanaceae]

Matthiola livida (Delile) DC. [Cruciferae]
Moltkiopsis ciliata (Forssk.) I.M. Johnston [Boraginaceae]

Neurada procumbens L. [Rosaceae]
Nitraria retusa (Forssk.) Ascherson [Zygophyllaceae]
Nitraria schoberi L. [Zygophyllaceae]
Noaea mucronata (Forssk.) Ascherson et Schweinf. [Chenopodiaceae]

Oenothera californica (S. Watson) S. Watson subsp. *eurekensis* (Munz et Roos) W. Klein [Onagraceae]
= *Oenothera ovita* subsp. *eurekensis* (see *O. californica*)
Ononis natrix L. [Papilionaceae]
Ononis serrata Forssk. [Papilionaceae]
Ononis variegata L. [Papilionaceae]
Orobanche cernua Loefl. [Scrophulariaceae]
= *Oryzopsis hymenoides* (Roemer et Schultes) Ricker (see *Achnatherum*)

Palafoxia arida B. Turner et M. Noris [Asteraceae]
Pancratium maritimum L. [Amaryllidaceae]
Pancratium sickenbergeri C. et W. Barbey [Amaryllidaceae]
Panicum turgidum Forssk. [Poaceae]
Panicum urvilleanum Kunth. [Poaceae]
Papaver humile Fedde [Papaveraceae]
Paronychia arabica (L.) DC. [Caryophyllaceae]
Pennisetum ciliare (L.) Link [Poaceae]
= *Pennisetum dichotomum* (Forssk.) Delile (see *P. divisum*)
Pennisetum divisum (J.F. Gmelin) Henrard [Poaceae]
Pennisetum elatum Hochst. et Steud. [Poaceae]
Petalonyx thurberi Gray [Loasaceae]
Phoenix dactylifera L. [Palmae]
Phragmites australis (Cav.) Steud. [Poaceae]
Picris asplenioides L. [Asteraceae]
= *Picris radicata* (Forssk.) Less. (see *P. asplenioides*)
Pistacia lentiscus L. [Anacardiaceae]
Pistacia terebinthus L. [Anacardiaceae]
Plantago cylindrica Forssk. [Plantaginaceae]
Plantago ovata Forssk. [Plantaginaceae]
Polycarpon succulentum (Delile) Gay [Caryophyllaceae]
Polygonum palaestinum Zohary [Polygonaceae]
= *Poterium spinosum* L. (see *Sarcopoterium*)
Prosopis farcta (Banks et Solander) Macbride [Mimosaceae]
Prosopis glandulosa Torrey [Mimosaceae]
Prunus maritima Wangenh. [Rosaceae]
Prunus serotina Ehrh. [Rosaceae]
Pseudorlaya pumila (L.) Grande [Umbelliferae]
Psorothamnus polyadenius Rydb. [Papilionaceae]
Pteranthus dichotomus Forssk. [Caryophyllaceae]

Retama raetam (Forssk.) Webb [Papilionaceae]
Rhamnus lycioides L. [Rhamnaceae]

Rostraria smyrnacea (Trin.) H. Scholz [Poaceae]
Rumex pictus Forssk. [Polygonaceae]

Salsola nollothensis Aellen [Chenopodiaceae]
Salsola tetrandra Forssk. [Chenopodiaceae]
Salvadora persica L. [Salvadoraceae]
Sarcobatus vermiculatus (Hooker) Torrey [Chenopodiaceae]
Sarcopoterium spinosum (L.) Spach [Rosaceae]
Satureja thymbra L. [Lamiaceae]
Savignia parviflora (Delile) Webb [Cruciferae]
Scabiosa eremophila Boiss. [Dipsacaceae]
Schimpera arabica Hochst. et Steudel [Cruciferae]
Schismus arabicus Nees [Poaceae]
Scirpus holoschoenus L. [Cyperaceae]
Scrophularia hypericifolia Wydler [Scrophulariaceae]
Senecio glaucus L. [Asteraceae]
Senecio joppensis Dinsmore [Asteraceae]
Silene colorata Poiret [Caryophyllaceae]
Silene sedoides Poiret [Caryophyllaceae]
Silene succulenta Forssk. [Caryophyllaceae]
Silene villosa Forssk. [Caryophyllaceae]
Spartina patens Muhl. [Poaceae]
= *Sporobolus arenarius* (Gouan) Duv.-Jouve (see S. *pungens*)
Sporobolus pungens (Schreber) Kunth [Poaceae]
Stipa capensis Thunb. [Poaceae]
Stipa comata Trin. et Rupr. [Poaceae]
= *Stipa hymenoides* Roemer et Schultes (see *Achnatherum*)
Stipa pellita (Trin. et Rupr.) Tzvelev [Poaceae]
Stipagrostis acutiflora (Trin. et Rupr.) de Winter [Poaceae]
Stipagrostis amabilis (Schweickerst) de Winter [Poaceae]
Stipagrostis ciliata (Desf.) de Winter [Poaceae]
Stipagrostis drarii (Taeckholm) de Winter [Poaceae]
Stipagrostis hirtigluma (Trin. et Rupr.) de Winter [Poaceae]
Stipagrostis karellinii (Trin. et Rupr.) Scholz [Poaceae]
Stipagrostis lanata (Forssk.) de Winter [Poaceae]
Stipagrostis lutescens (Nees) de Winter [Poaceae]
Stipagrostis namibensis de Winter [Poaceae]
Stipagrostis obtusa (Delile) Nees [Poaceae]
Stipagrostis pennata (Trin.) de Winter [Poaceae]
Stipagrostis plumosa (L.) T. Andersson [Poaceae]
Stipagrostis pungens (Desf.) de Winter [Poaceae]
Stipagrostis raddiana (Savi) de Winter [Poaceae]
Stipagrostis sabulicola (Pilger) de Winter [Poaceae]
Stipagrostis scoparia (Trin. et Rupr.) de Winter [Poaceae]
Stipagrostis seelyae de Winter [Poaceae]
Suaeda vermiculata J.F. Gmel. [Chenopodiaceae]
Swallenia alexandrae (Swallen) Soederstron & Decker [Poaceae]

Tamarix aphylla (L.) Karsten [Tamaricaceae]
Tamarix ramosissima Ledeb. [Tamaricaceae]
Tetradymia tetrameres (S.F. Blake) Strother [Asteraceae]
Thymelaea hirsuta (L.) Endl. [Thymelaeaceae]
= *Thymus capitatus* L. (see *Coridothymus)*
Tiquilia plicata (Torrey) A. Richardson [Boraginaceae]
Tiquilia nuttalii (Bentham) A. Richardson [Boraginaceae]
Traganum nudatum Delile [Chenopodiaceae]
Trianthema hereroensis Schniz [Aizooaceae]
Trifolium tomentosum L. [Papilionaceae]
Trigonella arabica Delile [Papilionaceae]
Trisetaria linearis Forssk. [Poaceae]

Uniola paniculata L. [Poaceae]

Vitis vinifera L. [Vitaceae]
Vulpia brevis Boiss. et Kotschy [Poaceae]

Zygophyllum album L.f [Zygophyllaceae]
Zygophyllum dumosum Boiss. [Zygophyllaceae]

References

Abeliovich A, Shilo M (1972) Photooxidative death in blue-green algae. J Bacteriol 111:682

Akers TG, Edmonds RL, Krame CL, Lighthart B, McManus ML, Schlichting HE Jr, Solomon AM, Splendlove JC (1979) Sources and characteristics of airborne materials. In: Edmonds RL (ed) Aerobiology. Dowden, Hutchinson and Ross, Stroudsburg, pp 11–84

Ancker van den JAM, Jungerius PD, Mur LR (1985) The role of algae in the stabilization of coastal dune blowouts. Earth Surf Proc Landf 10:189–192

Anderson DC, Harper KT, Holmgren RC (1982) Factors influencing development of cryptogamic crusts in Utah deserts. J Range Manage 35:180–185

Arber A (1934) The Gramineae. Cambridge University Press, London

Ash JE, Wasson J (1983) Vegetation and sand mobility in the Australian desert dunefield. Z Geomorphol Suppl 45:7–25

Bagnold RA (1941) The physics of blown sand and desert dunes. Methuen, London

Baierle HU (1993) Vegetation und Flora im südwestlichen Jordanien. Dissertationes Botanicae 200, Cramer, Berlin

Bailey D, Mazurak AP, Rosowski JR (1973) Aggregation of soil particles by algae. J Phycol 9:99–101

Barbour MG (1969) Age and space distribution of the desert shrub *Larrea divaricata*. Ecology 50: 679–685

Barbour MG, Craig RB, Drysdale FR, Ghiselin MT (1973) Coastal ecology, Bodega Head. University of California Press, Berkeley

Barbour MG, DeJong TM, Pavlik BM (1985) Marine beach and dune plant communities. In: Chabot BF, Mooney HA (eds) Physiological ecology of North American plant communities. Chapman and Hall, New York, pp 296–322

Barbour MG, Burk JH, Pitts WD (1987) Terrestrial plant ecology, 2nd edn. Benjamin and Cummings, Menlo Park

Barbour MG, Pavlik BM, Drysdale FR, Lindstrom S (1993) California's changing landscapes. California Native Plant Society, Sacramento

Barlow PW (1986) Adventitious roots of whole plants: their forms, functions, and evolution. In: Jackson MB (ed) New root formation in plants and cuttings. Martinus Nijhoff, Dordrecht, pp 67–110

Barlow PW (1994) The origin, diversity and biology of shoot-borne roots. In: Davis TD, Haissig BE (eds) Biology of adventitious root formation. Plenum, New York, pp 1–23

Bar-Or Y, Danin A (1989) Mechanisms of cyanobacterial crust formation and soil stabilization in the northwestern Negev. Isr Soc Microbiol, Haifa, p 13

Bar-Or Y, Shilo M (1987) Characterization of macromolecular flocculants produced by *Phormidium* sp. strain J-1 and by *Anabaenopsis circularis* PCC 6720. Appl Environ Microbiol 53:2226–2230

Bar-Or Y, Shilo M (1988) The role of cell-bound flocculants in coflocculation of benthic cyanobacteria with clay particles. FEMS Microbiol Ecol 53:169–174

Barrere P (1992) Dynamics and management of the coastal dunes of the Landes, Gascony, France. In: Carter RWG, Curtis TGF, Sheehy-Skeffington MJ (eds) Coastal dunes. Balkema, Rotterdam, pp 25–32

Batanouny KH, Batanouny MH (1968) Formation of phytogenic hillocks. I. Plants forming phytogenic hillocks. Acta Bot Acad Sci Hung 14:243–252

Batanouny KH, Batanouny MH (1969) Formation of phytogenic hillocks. II. Rooting habits of plants forming phytogenic hillocks. Acta Bot Acad Sci Hung 15:1–18
Bell A (1991) Plant morphology. Oxford University Press, Oxford
Bendali F (1987) Dynamique de la végétation et mobilité du sable en Jeffara Tunisien. PhD Thesis, Université de Montplellier Inst Reg Arides, Tunisie, 243 pp
Bendali F, Floret C, Le Floc'h E, Pontanier R (1990) The dynamics of vegetation and sand mobility in arid regions of Tunisia. J Arid Environ 18:21–32
Billings WD (1945) The plant associations of the Carson desert region, western Nevada. Butler Univ Bot Stud 8:89–123
Binyamini N (1980) Addenda to the hypogeous mycoflora of Israel. Nova Hedwigia 32:9–20
Boissier PE (1879) Flora Orientalis. H Georg, Basel, Geneve and Lyon 4:997–1001
Booth WE (1941) Algae as pioneers in plant succession and their importance in erosion control. Ecology 22:38–46
Bor NL (1970) Graminae. In: Rechinger KH (ed) Flora Iranica No 70. Academischer Druck, Graz
Boyce SG (1954) The salt spray community. Ecol Monogr 24:29–67
Bradshaw AD, Hardwick K (1989) Evolution and stress – genotypic and phenotypic components. Biol J Linn Soc 37:137–155
Brandon RN (1990) Adaptation and environment. Princeton University Press, Princeton
Breckon GJ, Barbour MG (1974) Review of North American Pacific coast beach vegetation. Madrono 22:333–360
Brock TD (1975) Effect of water potential on a *Microcoleus* (Cyanophyceae) from a desert crust. J Phycol 11: 316–320
Buckley RC (1983) Soil nitrogen requirements of tropical sandridge plants. Biotropica 15:77–78
Buckley RC, Chen W, Liu Y, Zhu Z (1986) Characteristics of the Tengger dunefield north-central China and comparison with the central Australian dunefields. J Arid Environ 10:97–102
Burns MER, Lubke RA (1986) Succession and zonation in the East London dune forest. Naturalist 30:38–44
Buzer JS, Dohmeier RA, Toit du DR (1985) The survival of algae in dry soils exposed to high temperatures for extended time periods. Phycologia 24:249–251
Campbell RS (1929) Vegetative succession in the *Prosopis* sand dunes of southern New Mexico. Ecology 10:392–398
Carlquist S (1974) Island biology. Columbia University Press, New York
Chadwick HW, Dalke P (1965) Plant succession on dune sands in Fremont County Idaho. Ecology 46:765–780
Chapman VJ (1976) Coastal vegetation, 2nd edn. Pergamon Press, Oxford
Chatigny MA, Dimmick RL, Mason CJ (1979) Atmospheric transport. In: Edmonds RL (ed) Aerobiology. Dowden, Hutchinson and Ross, Stroudsburg, pp 85–109
Conrad M (1983) Adaptability. Plenum Press, New York
Cooke RU, Warren A (1973) Desert geomorphology. Batsford, London
Cooper WS (1936) The strand and dune flora of the Pacific coast of North America: a geographic study. In: Goodspeed TH (ed) Essays in geobotany. University California Press, Berkeley, pp 141–187
Cooper WS (1958) Coastal sand dunes of Oregon and Washington. Geol Soc Am Mem 72:1–168
Corre JJ (1975) Étude d'un massif dunaire le long du littoral méditerranéen, structure et dynamisme du milieu et de la végétation. In: Gehu JM (ed) La Végétation des Dunes Maritimes Paris (1971). Cramer, Vaduz, pp 201–224
Crawford RMM (1989) Studies in plant survival. Blackwell, Oxford
Dafni A (1975) Comparative biology of *Prosopis farcta* (Banks et Sol.) Macbride, from different habitats in Israel. PhD Thesis, The Hebrew University of Jerusalem, Israel (in Hebrew with English Abstr)
Danin A (1964) Vegetation of Mishor Rotem-Yamin, Israel. The Hebrew University of Jerusalem, Israel (unpublished MSc Thesis, in Hebrew)
Danin A (1974) Notes on the vegetation near Suez and Fayid (Egypt). Isr J Bot 23:226–236

Danin A (1976) *Astragalus camelorum* Barbey, a rediscovered species from the Isthmic Desert (N Sinai). Isr J Bot 24:214–215

Danin A (1978a) Species diversity of semishrub xerohalophyte communities in the Judean Desert of Israel. Isr J Bot 27:66–76

Danin A (1978b) Plant species diversity and plant succession in a sandy area in the northern Negev. Flora 167:409–422

Danin A (1981a) The impact of geomorphologic and climatic conditions on the vegetation of the salt marshes along the Mediterranean coast of Israel and Sinai. Actas III Congress OPTIMA Anal Jar Bot Madrid 37:269–275

Danin A (1981b) Ecological factors affecting the distribution of *Tamarix aphylla* (L.) Karsten in Israel and Sinai. La-Yaaran 31:8–11, 46–48

Danin A (1983) Desert vegetation of Israel and Sinai. Cana, Jerusalem

Danin A (1986) Patterns of biogenic weathering as indicators of palaeoclimates in Israel. Proc Roy Soc Edinb Biol Ser 89B:243–253

Danin A (1987) Impact of man on biological components of desert ecosystems in Israel. Isr J Bot 36:46

Danin A (1991) Plant adaptation to desert dunes. J Arid Environ 21:193–212

Danin A (1994) Adaptations of *Stipagrostis* species to desert dunes. Int Symp IAVS Bern, Switzerland MO/2A

Danin A (1995a) Man and the natural environment. In: Levy TE (ed) The archaeology of society in the Holy Land. Leicester University Press, London, pp 24–39

Danin A (1995b) A new *Anchusa* from Israel. Edinburgh J. Bot. 52(3):333–336

Danin A, Ganor E (1991) Trapping of airborne dust by mosses in the Negev desert, Israel. Earth Surf Proc Landf 16:153–162

Danin A, Kukkonen I (1995) Contributions to the flora of Israel. VIII. A new *Cyperus* from Israel, *Cyperus sharonensis* Danin et Kukkonen sp. n. Isr J Plant Sci 43:77–82

Danin A, Nokrian R (1991) Dynamics of dune vegetation in the southern coastal area of Israel since 1945. Doc Phytosociol 13:281–296

Danin A, Orshan G (1990) The distribution of Raunkiaer's life forms in Israel as related to environment. J Veg Sci 1:41–48

Danin A, Orshan G (1995) Circular arrangement of *Stipagrostis ciliata* clumps in the Negev, Israel and near Gokaeb, Namibia. J Arid Environ 30:307–313

Danin A, Yaalon DH (1981) Trapping of silt by lichens and mosses in the desert environment of the Dead Sea region, Israel. Int Conf Aridic Soils Jerusalem, Israel, p 29

Danin A, Yaalon DH (1982) Silt plus clay sedimentation and decalcification during plant succession in sands of the Mediterranean coastal area of Israel. Isr J Earth-Sci 31:101–109

Danin A, Orshan G, Zohary M (1964) Vegetation of the Neogene sandy areas of the northern Negev. Isr J Bot 13:208–233

Danin A, Orshan G, Zohary M (1975) The vegetation of the northern Negev and the Judean Desert of Israel. Isr J Bot 24:118–172

Danin A, Bar-Or Y, Dor I, Yisraeli T (1989) The role of cyanobacteria in stabilization of sand dunes in southern Israel. Ecol Mediterr 15:55–64

Danin A, Barbour MG, Rae SP (1995) Plant succession in dunes of Bodega Head 1955–1995. (unpublished)

Darwin C (1859) On the origin of species. John Murray, London

DeDecker M (1976) The Eureka dunes. Fremontia 3:17–20

De Malach Y, Kagan-Zur V, Roth-Bejerano N, Danin A (1995) Symbiotic association between hypogeous mycoflora and *Helianthemum* species in Israel. (unpublished)

Disraeli DJ (1984) The effect of sand deposits on the growth and morphology of *Ammophila breviligulata*. J Ecol 72:145–154

Doing H (1986) Relationships between vegetation and geomorphology of coastal sand dunes in the Netherlands. In: Gehu JM (ed) Vegetation et geomorphologie. Cramer, Berlin, pp 411–429

Dor I (1995) Investigation of microphytic crusts in the desert soils of Israel. Annu Rep, The Hebrew University of Jerusalem, Israel, pp 1–32

Du Rietz GE (1931) Life-forms of terrestrial flowering plants. Acta Phytogeogr Suec III:1–95

Eig A (1939) The vegetation of the light soils belt of the Coastal Plain of Palestine. Palest J Bot Jerus Ser 1:255–308

Esau (1977) Anatomy of seed plants, 2nd edn. Wiley, New York

Evenari M, Shanan L, Tadmor N (1982) The Negev, the challenge of a desert. 2nd edn. Harvard University Press, Cambridge

Evenari M, Gutterman Y, Gavish E (1985) Botanical studies on coastal salines and sabkhas of the Sinai. In: Friedman GM, Krumbein WE (eds) Hypersaline Ecosystems, the Gavish sabkha. Ecological Studies 53. Springer, Berlin Heidelberg New York, pp 145–182

Fahn A (1982) Plant anatomy, 3rd edn. Pergamon Press, Oxford

Fahn A, Cutler D (1992) Xerophytes. Borntraeger, Berlin

Fahn A, Sarnat A (1963) Xylem structure and annual rythm of development in trees and shrubs of the desert. Bull Res Counc Isr 11(D):198–210

Feinbrun-Dothan N (1986) Flora Palaestina, part 4. Israel Acad Sci Humanities, Jerusalem

Feinbrun-Dothan N, Danin A (1991) Analytical flora of Eretz Israel. Cana, Jerusalem

Fireman M, Hayward HE (1952) Indicator significance of some shrubs in the Escalente Desert, Utah. Bot Gaz 114:143–155

Fletcher JE, Martin WP (1948) Some effects of algae and molds in the rain-crust of desert soils. Ecology 29:95–100

Flowers S (1934) Vegetation of the Great Salt Lake region. Bot Gaz 95:353–418

Fonteyn PJ, Mahall BE (1978) Competition among desert perennials. Nature 275:544–545

Friedmann EI, Galun M (1974) Desert algae, lichens, and fungi. In: Brown GW (ed) Desert biology II. Academic Press, New York, pp 165–212

Fryberger SG (1991) Unusual sedimentary structures in the Oregon coastal dunes. J Arid Environ 21:131–150

Fuller WH, Cameron RE, Raica N (1960) Fixation of nitrogen in desert soils by algae. Working Pap 7th Congr Int Soc Soil Sci 19 Aug Madison, Wisconsin

Futuyma DJ (1986) Evolutionary biology. Sinauer, Massachusetts

Gabai VL (1985) A one-instant mechanism of phototaxis in the cyanobacterium *Phormidium unicinatum*. Fed Eur Microbiol Soc Microbiol Lett 30:125–129

Gehu JM (1985) European dune and shorline vegetation. Council of Europe, Nature and Environment Series No 32, Strasbourg

Gemmell AR, Greig-Smith P, Gimingham CH (1953) A note on the behavior of *Ammophila arenaria* (L.) Link in relation to sand dune formation. Trans Proc Bot Soc Edinb 36:132–136

Gimingham CH (1948) The role of *Barbula fallax* Hedw. and *Bryum pendulum* Schp. in sand-dune fixation. Trans Br Bryol Soc 1:70–72

Gimingham CH (1964) Maritime and sub-maritime communities. In: Burnett JH (ed) The vegetation of Scotland. Oliver and Boyd, Edinburgh, pp 67–125

Ginzburg C (1963) Some anatomical features of splitting of desert shrubs. Phytomorphology 13:92–97

Goring-Morris N, Goldberg P (1990) Late Quaternary dune incrusions in the southern Levant: archaeology, chronology and palaeoenvironments. Quat Int 5:115–137

Grant V (1963) The origin of adaptations. Columbia University Press, New York

Greeley R, Arvidson RE (1990) Aeolian processes on Venus. Earth Moon Planets 50/51:127–157

Greig-Smith P, Gemmell AR, Gimingham CH (1947) Tussock formation in *Ammophila arenaria* (L.) Link. New Phytol 46:262–268

Groff PA, Kaplan DR (1988) The relation of root systems to shoot systems in vascular plants. Bot Rev 54:387–422

Gutterman Y (1987) Dynamics of porcupine (*Hystrix indica* Kerr) diggings: their role in the survival and renewal of geophytes and hemicryptophytes in the Negev desert highlands. Isr J Bot 36:133–143

Gutterman Y (1993) Seed germination in desert plants. In: Cloudsley-Thompson JL (ed) Adaptations of desert organisms. Springer, Berlin Heidelberg New York

Haissig BE, Davis TD (1994) A historical evalustion of adventitious rooting research to 1993. In: Davis TD, Haissig BE (eds) Biology of adventitious root formation. Plenum Press, New York, pp 275–331

Halle F, Oldeman RAA, Tolinson PB (1978) Tropical trees and forests. Springer, Berlin Heidelberg New York

Harper JL (1977) Population biology of plants. Academic Press, London

Harper KT, Marble JR (1988) A role of nonvascular plants in management of arid and semiarid rangelands. In: Tueller PT (ed) Vegetation science applications for rangeland analysis and management. Kluwer, Dordrecht, pp 135–169

Harper KT, Pendleton RL (1993) Cyanobacteria and cyanolichens: can they enhance availability of essential minerals for higher plants? Great Basin Nat 53:59–72

Hastenrath S (1987) The barchan dunes of southern Peru revisited. Z Geomorphol NF 31:167–178

Henrard JT (1929) A monograph of the genus *Aristida*. Mede Rijks Herb Leiden 1:30–85

Hesp PA (1981) The formation of shadow dunes. J Sediment Petrol 51:101–112

Hickman JC (ed) (1993) The Jepson manual: higher plants of California. University of California Press, Berkeley

Hoerner SF (1965) Fluid-dynamic drag. SF Hoerner, Midland Park, New Jersey

Huiskes AHL (1979) Biological flora of the British Isles: *Ammophila arenaria* (L.) Link. J Ecol 67:363–382

Jackson R (1973) Diurnal changes in soil water content during drying. In: Field soil water regime. Soil Sci Soc Am, Spec Publ No 5, Madison, Wisconsin

Jackson RM, Mason PA (1984) Mycorrhiza. Edward Arnold, London

Johansen JR (1993) Cryptogamic crusts of semiarid and arid lands of North America. J Phycol 29:140–147

Johnson AF (1977) A survey of the strand and dune vegetation along the Pacific and southern Gulf coasts of Baja California. J Biogeogr 7:83–99

Jungerius PD, van der Meulen F, Loedeman JH, Stuiver J (1992) A geometrical approach to monitoring blowout development from aerial photographs using a geographical information system (GIS). In: Carter RWG, Curtis TGF, Sheehy-Skeffington MJ (eds) Coastal dunes. Balkema, Rotterdam, pp 129–138

Karnieli A, Tsoar H (1995) Spectral reflectance of biogenic crust developed on desert dunesand along the Israel–Egypt border. Int J Remote Sens 16:369–374

Keshet M, Danin A, Orshan G (1990) Distribution of ecomorphological types along environmental gradients in Israel. 1. Renewal bud location and leaf attributes. Ecol Mediterr 16: 151–161

Kind RJ (1990) Mechanics of aeolian transport of snow and sand. J Wind Eng Ind Aerodyn 36:855–866

Kleiner EF, Harper KT (1977) Soil properties in relation to cryptogamic ground cover in Canyonlands National Park. J Range Manage 30:202–205

Koller D (1956) Germination regulating mechanisms in some desert seeds. III.*Calligonum comosum* L'Her. Ecology 37:430–433

Koller D, Sachs M, Negbi M (1964) Germination regulating mechanisms in some desert seeds. VIII *Artemisia monosperma*. Plant Cell Physiol 5:85–100

Komarov VL (ed) (1936) Flora URSS. Academiae Scientiarum URSS, Leningrad Moscow 5:527–594

Koske RE (1975) *Endogene* spores in Australian sand dunes. Can J Bot 53:668–672

Koske RE, Halvorson WL (1981) Ecological studies of vascular-arbuscular mycorrhizae in a barrier sand dune. Can J Bot 59:1413–1422

Kramer PJ (1969) Plant and soil water relationships, 2nd edn. McGraw-Hill, New York

Kukkonen I (1994) Definition of descriptive terms for the Cyperaceae. Ann Bot Fenn 31:37–43

Kutiel P, Danin A, Orshan G (1979/80) Vegetation of the sandy soils near Caesarea, Israel. I. Plant succession. Isr J Bot 28:20–35

Lancaster N (1993) Kelso dunes. Nat Geogr Res Explor 9:444–459
Lange OL, Kidron GJ, Büdel B, Meyer A, Kilian E, Abeliovich A (1990) Taxonomic composition and photosynthetic characteristics of the 'biological soil crusts' covering sand dunes in the western Negev desert. Funct Ecol 6:519–527
Lange OL, Meyer A, Büdel B (1994a) Net photosynthesis activation of a desiccated cyanobacterium without liquid water in high air humidity alone. Experiments with *Microcoleus sociatus* isolated from a desert soil crust. Funct Ecol 8:52–57
Lange OL, Meyer A, Zellner H, Heber U (1994b) Photosynthesis and water relations of lichen soil crusts: field measurments in coastal fog zone of the Namib Desert. Funct Ecol 8:253–264
Lange W (1974) Chelating agents and blue-green algae. Can J Microbiol 20:1311–1321
Lapedes DN (ed) (1976) McGraw-Hill dictionary of the life sciences. McGraw-Hill, New York
Leistner OA (1967) The plant ecology of southern Kalahari. Bot Surv S Afr Mem 38
Leonard J (1981) Contribution a l'etude de la flora et de la végétation des déserts d'Iran. Jard Bot Belg, Meise 1:79
Levy Y, Steinberger Y (1986) Adaptation of desert soil microalgae to varying light intensities. Comp Physiol Ecol 11:90–94
Loeftis SG, Kurtz EB (1980) Field studies of inorganic nitrogen added to semiarid soils by rainfall and blue-green algae. Soil Sci 129:150–155
Louw GN, Seely MK (1982) Ecology of desert organisms. Longman, London
Lubke RA, Gess FW, Bruton MN (eds) (1988) A field guide to the eastern Cape Coast. The Grahamstown Centre of the Wildlife Society of Southern Africa, Grahamstown
Lyshede OB (1977) Anatomical features of some stem assimilating desert plants of Israel. Bot Tidsskr 71:225–230
Lyshede OB (1979) Xeromorphic features of some stem assimilants in relation to their ecology. J Linn Soc Lond Bot 78:85–98
Marshall JK (1965) *Corynephorus canescens* (L.) P. Beauv. as a model for the *Ammophila problem.* J Ecol 53:447–463
Marshall TJ, Holmes JW (1979) Soil physics. Cambridge University Press, New York
Marthe K (1972) Role of some blue-green algae in soil aggregation. In: Desikachary TV (ed) Taxonomy and biology of blue-green algae. University of Madras, Bangalore Press, Madras, pp 328–331
Mauseth JD (1988) Plant anatomy. Benjamin and Cummings, Menlo Park
Mayland HF, MacIntosh TH, Fuller WH (1966) Fixation of isotopic nitrogen in a semi-arid soil by algal crust otganisms. Soil Sci Am Proc 30:56–60
McClure FA (1966) The bamboos, a fresh perspective. Harvard University Press, Cambridge
McCully ME, Canny MJ (1989) Pathways and processes of water and nutrient movement in roots. In: Loughman BC, Gasparikova O, Kolek J (eds) Structural and functional aspects of transport in roots. Kluwer, Dordrecht, pp 3–14
McKee ED (1979) Introduction to a study of global sand seas. In: McKee ED (ed) A study of global sand seas. Geol Surv Prof Paper 1052, US Government, Washington, DC, pp 1–19
McKee ED (1982) Eolian sand bodies of the world. Proc Int Symp Remote Sens Environ 1: 75–82
Metcalff CR (1960) Anatomy of the monocotyledons. I. Graminae. Clarendon Press, London
Monod T (1931) Remarques biologiques sur la Sahara. Rev Gen Sci Pures Appl 42:609–616
Moore CJ, Scott GAM (1979) The ecology of mosses on a sand dune in Victoria, Australia. J Bryol 10:291–311
Mosse B (1973) Advances in the study of vesicular-arbuscular mycorrhiza. Annu Rev Phytopathol 11:171–196
Munz PA, Keck DD (1963) A California flora. University of California Press, Berkeley
Murray DR (1986) Seed dispersal. Academic Press, Sydney
Navarini L, Cesaro A, Ross-Murphy SB (1992) Viscoelastic properties of aqueous solutions of an exocellulr polysaccharide from Caynobacteria. Carbohydr Polymers 18:265–272
Newell RE (1971) The global circulation of atmospheric pollutants. Sci Am 224:32–42
Nienow JA, McKay CP, Friedmann EI (1988) The cryptoendolithic microbial environment in the Ross Desert of Antarctica: light in the photosynthetically active region. Microb Ecol 16:271–289

Niklas KJ (1992) Plant biomechanics. University Chicago Press, Chicago
Nokrian R (1988) Vegetation and its habitats in sands south of Nahal Shiqma, Israel. The Hebrew University of Jerusalem, Israel (unpublished MSc Thesis, in Hebrew)
Norris RM (1966) Barchan dunes dunes of Imperial Valley, California. Am J Geol 74:292-306
Noy-Meir I (1973) Desert ecosystems: environment and producers. Annu Rev Ecol Syst 4: 25-51
Noy-Meir I, Seligman NG (1979) Management of semi-desert ecosystems in Israel. In: Walker BH (ed) Management of semi-desert ecosystems. Elsevier, Amsterdam, pp 113-160
Olson JS (1958a) Lake Michigen dune development. II Plants as agents and tools in geomorphology. J Geol 66:346-351
Olson JS (1958b) Rates of succession and soil changes on southern Lake Michigen sand dunes. Bot Gaz 119:125-170
Oosting HJ, Billings DW (1942) Factors affecting vegetation zonation on coastal dunes. Ecology 23:131-142
Orshan G (1986a) The deserts of the Middle East. In: Evenari M, Noy-Meir I, Goodall DW (eds) Ecosystems of the world, vol 12B. Hot deserts and arid shrublands. Elsevier, Amsterdam, pp 1-28
Orshan G (1986b) Plant form as describing vegetation and expressing adaptation to environment. Ann Bot (Rome) 44:7-38
Orshan G, Zohary M (1963) Vegetation of the sand deserts in the western Negev of Israel. Vegetatio 11:112-120
Pacioni G, La Salvia M, Rosa S, Trotta G (1985) The annual dynamic of VA-mycorrhiza in a maritime sandy dune (central east Italy). In: Molina R (ed) 6th North American Conference on Mycorrhizae, Forest Res Lab, Corvallis, p 286
Pavlik BM (1980) Patterns of water potential and photosynthesis of desert sand dune plant, Eureka Valley, California. Oecologia (Berl) 46:147-154
Pavlik BM (1985) Sand dune flora of Great Basin and Mojave deserts of California, Nevada, and Oregon. Madroño 32:197-213
Pavlik BM, Barbour MG (1988) Demographic monitoring of endemic sand dune plants, Eureka Valley, California. Biol Conserv 46:217-242
Pijl van der L (1972) Principles of dispersal in higher plants. Springer, Berlin Heidelberg New York
Price RS (1911) The roots of some North African desert grasses. New Phytol 10:328-339
Prill RC (1968) Movement of moisture in the unsaturated zone in a dune area, southwestern Kansas. US Geol Surv Prof Paper 600D:1-9
Puppi G, Tabacchini P, Chiapperi F, Carpigo F (1985) Mycorrhiza in litoral Mediterranean plant communities. In: Molina R (ed) 6th North American Conference on Mycorrhizae, Forest Res Lab, Corvallis, p 324
Pye K, Tsoar H (1990) Aeolian sand and sand dunes. Unwin Hyman, London
Ranwell DS (1972) Ecology of salt marshes and sand dunes. Chapman and Hall, London
Raven PH, Evert RH, Curtis H (1986) Biology of plants. 4th edn. Worth, New York
Rendell HM, Yair A, Tsoar H (1993) Thermoluminescence dating of periods of sand movement and linear dune formation in the northern Negev, Israel. In: Pye K (ed) The dynamics and environment context of aeolian sedimentary systems. Geol Soc Spec Publ 72:69-74
Ries RE, Hofmann L (1991) Research observations: Standardized terminology for structures resulting in emergence and crown placement of 3 perennial grasses. J Range Manage 44:404-407
Robertson JH (1983) Greaswood [*Sarcobatus vermiculatus* (Hook.) Torr.]. Phytologia 54:309-324
Robinson TW (1958) Phreatophytes. US Geol Surv Water Supply Pap 1423, pp 1-84
Rollins MB, Dylla AS, Eckert RE Jr (1968) Soil problems in reseeding a greasewood - rabbitbrush range site. J Soil Water Conserv 23:131-140
Rosenan N, Gilead A (1985) Maps of mean annual precipitation and temperature. Atlas of Israel, 3rd edn. Dept Surveys, Ministry of Labour, Jerusalem, p 14/1
Rost TL, Barbour MG, Thornton RM, Weier TE, Stocking CR (1984) Botany, 2nd edn. Wiley, New York

Roth-Bejerano N, Livne D, Kagan-Zur V (1990) *Helianthemum-Terfezia* relations in different growth media. New Phytol 114:235–238

Rychert RC, Skujins J (1974) Nitrogen fixation by blue-green algae – lichen crusts in the Great Basin Desert. Soil Sci Soc Am Proc 38:768–771

Sabharwal A, Mukerji KG (1985) Occurrence and characteristics of VAM in ecologically different sites. In: Molina R (ed) 6th North American Conference on Mycorrhizae, Forest Res Lab, Corvallis, p 316

Sarnthein M (1978) Sand deserts during glacial maximum and climatic optimum. Nature 272: 43–46

Schofield WB (1985) Introduction to bryology. Macmillan, New York

Scott GAM (1982) Desert bryophytes. In: Smith AJE (ed) Bryophyte ecology. Chapman and Hall, London

Seely MK (1991) Sand dune communities. In: Polis GA (ed) The ecology of desert communities. University of Arizona Press, Tucson, pp 348–382

Seely MK, Louw GN (1980) First approximation of the effects of rainfall on the ecology and energetics of a Namib Desert dune ecosystem. J Arid Environ 3:25–54

Seliskar DM (1994) The effect of accelerated sand accretion on growth, carbohydrate reserves, and ethylene production in *Ammophila breviligulata* (Poaceae). Am J Bot 81:536–541

Sheldon JC, Burrows FM (1973) The dispersal effectiveness of the achene-pappus units of selected Compositae in steady winds with convection. New Phytol 72:663–675

Smith SSE (1980) Mycorrhizas of autotrophic higher plants. Biol Rev 55: 475–510

Starks TL, Shubert LE, Trainor FR (1981) Ecology of soil algae: a review. Phycologia 20:65–80

St. Clair LI, Johansen JR (1993) Introduction to the symposium on soil crust communities. Great Basin Nat 53:1–4

Stebbins GL (1971) Adaptive radiation of reproductive characteristics in angiosperms, II. Seeds and seedlings. Annu Rev Ecol Syst 2:237–260

Stewart WDP (1967) Transfer of biologically fixed nitrogen in a sand dune slack region. Nature 214:603–604

Sutton JC, Sheppard BR (1976) Aggregation of sand dune soil by endomycorrhizal fungi. Can J Bot 54:326–333

Sylvia DM (1985) Mycorrhizal fungi associated with sea oats in Florida sand dunes. In: Molina R (ed) 6th North American Conference on Mycorrhizae, Forest Res Lab, Corvallis, p 282

Thomas DSG (1992) Desert dune activity: concepts and significance. J Arid Environ 22:31–38

Thomas DSG, Tsoar H (1990) The geomorphological role of vegetation in desert dune systems. In: Thornes JB (ed) Vegetation and erosion. Wiley, New York

Tinley KL (1985) Coastal dunes of South Africa. S Afr Nat Sci Prog Rep 109, Pretoria

Troughton A (1957) The underground organs of herbage grasses. England Commonwealth Agricultural Bureau, Farnham Royal, Bucks

Tsoar H (1974) Desert dunes morphology and dynamics, El Arish (northern Sinai). Z Geomorphol NF Suppl Bd 20:41–61

Tsoar H (1982) Sand dunes and their modes of activity. Mada 26:13–19 (in Hebrew)

Tsoar H, Møller JT (1986) The role of vegetation in the formation of linear sand dunes. In: Nickling WG (ed) Aeolian geomorphology. Allen and Unwin, Boston, pp 74–95

Tsoar H, Pye K (1987) Dust transport and the question of desert loess formation. Sedimentology 34:139–153

Vaadia Y, Waisel Y (1963) Water absorption by aerial organs of plants. Physiol Plant 16: 44–51

Vasek FC (1980) Creosote bush: long-lived clones in the Mojave Desert. Am J Bot 67:246–255

Vermeer J, McCully ME (1982) The rhizosphere in Zea: new insight into its structure and development. Planta 156:45–61

Volkens G (1887) Die Flora der aegyptisch-arabischen Wueste auf Grundlage anatomisch-physiologischer Forschungen. Gebrueder Borntraeger, Berlin

Walter H (1973) Vegetation of the Earth. Springer, Berlin Heidelberg New York

Walter H, Box EO (1983a) The Karakum desert, an example of a well-studied eu-biome. In: West N (ed) Ecosystems of the world, vol 5. Temperate deserts and semideserts. Elsevier, Amsterdam, pp 105–159
Walter H, Box EO (1983b) The deserts of central Asia. In: West N (ed) Ecosystems of the world, vol 5. Temperate deserts and semideserts. Elsevier, Amsterdam, pp 193–236
Warming E (1909) Oecology of plants. Oxford University Press, London
Watt AS (1937) Studies in the ecology of Brekland. II. On the origin and development of blowouts. J Ecol 25:91–112
Weaver JE, Clements FE (1929) Plant ecology. McGraw-Hill, New York
West NE (1990) Structure and function of microphytic soil crusts in wildland ecosystems of arid and semi-arid regions. Adv Ecol Res 20:179–223
West CP (1994) Physiology and drought tolerance of endophyte-infected grasses. In: Bacon CW, White JF Jr (eds) Biotechnology of endophytic fungi of grasses. CRC Press, Boca Raton, pp 87–98
Wilson K, Honey JN (1966) Root contraction in *Hyacinthus orientalis*. Ann Bot 30:47–61
Winder de B, Pluis J, Reus de L, Mur LR (1989) Characterization of a cyanobacterial algal crust in the coastal dunes of the Netherlands In: Cohen Y, Rosenberg E (eds) Microbial mats, physiological ecology of benthic microbial communities. AMS, Washington, DC
Wullstein LH (1991) Variation in N_2 fixation (C_2H_2 reduction) associated with rhizosheaths of Indian Ricegrass (*Stipa hymenoides*). Am Midl Nat 126:76–81
Wullstein LH, Pratt SA (1981) Scanning electron microscopy of rhizosheaths of *Oryzopsis hymenoids*. Am J Bot 68:408–419
Wullstein LH, Bruening ML, Bollen WB (1979) Nitrogen fixation associated with sand grain root sheaths (rhizosheaths) of certain xeric grasses. Physiol Plant 46:1–4
Yaalon DH, Dan Y (1974) Accumulation and distribution of loess-derived deposits in the semidesert and desert fringe areas of Israel. Z Geomorphol NF 20:91–105
Yair A (1990) Runoff generation in a sandy area – the Nizzana sands, eastern Negev, Israel. Earth Surf Proc Landf 15:597–609
Yeaton RI (1988) Structure and function of the Namib dune grassland: characteristics of the environmental gradients and species distribution. J Ecol 76:744–758
Yeaton RI (1990) The structure and function of the Namib dune grasslands: species interactions. J Arid Environ 18:343–349
Zanten BO van (1978) Experimental studies on trans-oceanic long-range dispersal of moss spores in the Southern Hemisphere. J Hattori Bot Lab 44:455–482
Zanten BO van (1984) Some considerations on the feasibility of long-distance transport in bryophytes. Acta Bot Neerl 33:231–232
Zilberman E (1982) The geology of the Qeren–Haluza area. Isr Geol Surv Rep EG/4/1982, Jerusalem (in Hebrew)
Zilberman E (1989) Geomorphic relations between the Upper Pleistocene terraces of Nahal Lavan and the anticline of Qeren Ridges; possible evidence for neotectonic uplift. Isr Geol Surv Rep GSI/20/89, Jerusalem (in Hebrew)
Zohary M (1973) Geobotanical foundations of the Middle East. Fischer, Stuttgart
Zohary M, Fahn A (1952) Ecological studies on East Mediterranean dune plants. Bull Res Counc Isr Sect D 1:38–53

Subject Index

Plant Index

MIX
Papier aus verantwortungsvollen Quellen
Paper from responsible sources
FSC® C105338

FSC
www.fsc.org

If you have any concerns about our products,
you can contact us on
ProductSafety@springernature.com

In case Publisher is established outside the EU,
the EU authorized representative is:
Springer Nature Customer Service Center GmbH
Europaplatz 3, 69115 Heidelberg, Germany

Printed by Libri Plureos GmbH
in Hamburg, Germany